Ben Stacy Jerrik (Ed.)

Interstate 185 (South Carolina)

AF307106

Ben Stacy Jerrik (Ed.)

Part Press

Interstate 185 (South Carolina)

U.S. Route 29, Interstate 385, Interstate 85, South Carolina Department of Transportation

Part Press

Imprint

Permission is granted to copy, distribute and/or modify this document under the terms of the GNU Free Documentation License, Version 1.2 or any later version published by the Free Software Foundation; with no Invariant Sections, with the Front-Cover Texts, and with the Back- Cover Texts. A copy of the license is included in the section entitled "GNU Free Documentation License".

All parts of this book are extracted from Wikipedia, the free encyclopedia (www.wikipedia.org).

You can get detailed informations about the authors of this collection of articles at the end of this book. The editors (Ed.) of this book are no authors. They have not modified or extended the original texts.

Pictures published in this book can be under different licences than the GNU Free Documentation License. You can get detailed informations about the authors and licences of pictures at the end of this book.

The content of this book was generated collaboratively by volunteers. Please be advised that nothing found here has necessarily been reviewed by people with the expertise required to provide you with complete, accurate or reliable information. Some information in this book maybe misleading or wrong. The Publisher does not guarantee the validity of the information found here. If you need specific advice (f.e. in fields of medical, legal, financial, or risk management questions) please contact a professional who is licensed or knowledgeable in that area.

Any brand names and product names mentioned in this book are subject to trademark, brand or patent protection and are trademarks or registered trademarks of their respective holders. The use of brand names, product names, common names, trade names, product descriptions etc. even without a particular marking in this works is in no way to be construed to mean that such names may be regarded as unrestricted in respect of trademark and brand protection legislation and could thus be used by anyone.

Cover image: www.ingimage.com
Concerning the licence of the cover image please contact ingimage.

Publisher:
Part Press is a trademark of
International Book Market Service Ltd., 17 Rue Meldrum, Beau Bassin, 1713-01 Mauritius
Email: info@bookmarketservice.com
Website: www.bookmarketservice.com

Published in 2011

Printed in: U.S.A., U.K., Germany. This book was not produced in Mauritius.

ISBN: 978-613-8-91282-8

Contents

Articles

References

Interstate_185_(South_Carolina)

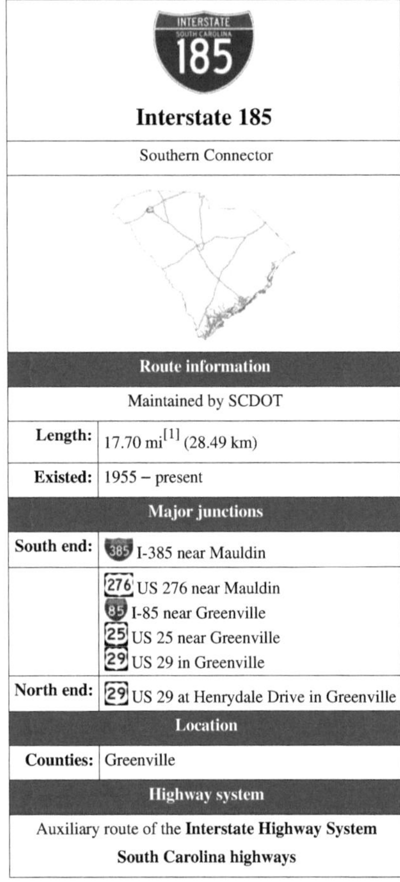

Interstate 185	
Southern Connector	
Route information	
Maintained by SCDOT	
Length:	17.70 mi[1] (28.49 km)
Existed:	1955 – present
Major junctions	
South end:	I-385 near Mauldin
	US 276 near Mauldin I-85 near Greenville US 25 near Greenville US 29 in Greenville
North end:	US 29 at Henrydale Drive in Greenville
Location	
Counties:	Greenville
Highway system	
Auxiliary route of the **Interstate Highway System** **South Carolina highways**	

Interstate 185 (I-185) is located in the city of Greenville, South Carolina. The northern portion, which ends just shy of the Greenville city limits, was opened in the 1960s and is cosigned with U.S. 29. The southern portion, which connects the I-85/I-185 interchange (exit 42) with the I-385/U.S. 276 interchange (exit 30), was opened as a toll road in 2001 ($1.25 per passenger vehicle). This extension was dubbed the "Southern Connector" and increased I-185 from three to seventeen miles (27 km) in length.

Southern Connector

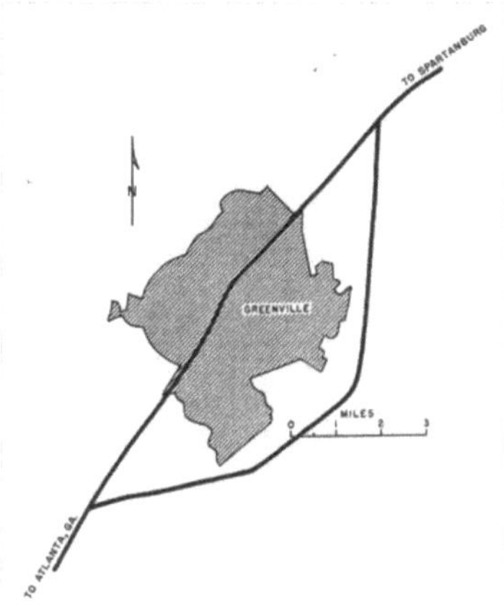

I-185 was planned as part of a continuous route through Greenville on this 1955 map. Note that I-85 would have used the US 29 corridor from Greenville north towards Spartanburg.

The Southern Connector was constructed as a public-private partnership between the South Carolina Department of Transportation and Interwest Carolina Transportation Group, LLC, a development team that included a not-for-profit corporation called Connector 2000 Association, Inc. (C2A).[2] Under this agreement, C2A operates the toll road under a fifty year license. They were responsible for financing, designing, constructing, operating and maintaining the road during this period and the toll revenue would be used to pay them for these efforts. To finance the project, C2A sold bonds that were tax-exempt under IRS Rule 63-20, which provides that the bonds sold will be exempt if they finance an activity which is "public in nature."[3] [4]

The highway opened in February 2001, nine months ahead of schedule.[5] By 2007, the Connector 2000 Association was having financial difficulties because ridership on the toll road was not meeting original estimates. In the fall of 2007, they began looking for a concessionaire to take over the operation and financial liability of the toll road.[6] By early 2008, C2A had received a default notice from their bond trustee[7] In January 2010, the bond trustee missed an interest payment,[8] and the C2A was more than $8 million behind in its payments to SCDOT for the maintenance and license fees under their agreement.[9] On June 24, 2010, the Southern Connector filed for bankruptcy.[10] Travelers on the Southern Connector can pay using electronic toll collection facilities. South Carolina uses the PalmettoPass,[11] which can be used on both the Southern Connector and the Cross Island Parkway on Hilton Head Island.

Exit list

The entire route is in Greenville County.

Location	Exit #	Destination	Notes
Mauldin		385 I-385 north – Greenville	Southbound exit and northbound entrance
	1C	417 SC 417 – Mauldin	Signed as exit 1 northbound; no northbound entrance to I-385 north
	1B	385 I-385 south – Simpsonville	Southbound exit and northbound entance
	1A	276 417 US 276 to SC 417 – Mauldin	Southbound exit and northbound entrance
Toll plaza			
	4	Fork Shoals Road	
	7	25 US 25 (Augusta Road) – Idlewild, Moonville	Signed as exits 7A (north) and 7B (south) northbound
Piedmont	10	20 SC 20 (Piedmont Highway) – Greenville	
Toll plaza			
	12	153 SC 153 west – Easley	
	14	85 29 I-85 / US 29 south – Spartanburg, Anderson	South end of US 29 overlap; signed as exits 14A (north) and 14B (south)

	15	[25] US 25 (White Horse Road) – Travelers Rest	
Greenville	16	Washington Avenue, Faris Road	
		Mills Avenue, Henrydale Avenue	At-grade intersection
		[29] US 29 north (Mills Avenue) – Greenville	Continuation beyond on Henrydale Avenue

References

[1] "Auxiliary Routes of the Dwight D. Eisenhower National System Of Interstate and Defense Highways as of October 31, 2002" (http://www.fhwa.dot.gov/reports/routefinder/table2.htm). Federal Highway Administration. .

[2] Samuel, Peter (October 26, 2007). "Greenville Southern Connector up for concession offers – not-for-profit ailing" (http://www.tollroadsnews.com/node/3211). TollRoadsNews. .

[3] "27-in-7 Peak Performance" (http://www.dot.state.sc.us/inside/pdfs/27in7.pdf) (PDF). SCDOT. .

[4] Hedlund, Karen J. (May 1, 2001). "The Use of '63-20' Nonprofit Corporations in Infrastructure Facility Development" (http://www.nossaman.com/showarticle.aspx?show=854). Nossaman Infrastructure. .

[5] "Partners unveil Southern Connector" (http://www.southernconnector.com/press04_Feb-16-01.htm) (Press release). Southern Connector Toll Road. February 16, 2001. .

[6] "Request for Toll Road Concessionaire Qualifications" (http://www.southernconnector.com/pdfs/SCTR_RFQ2.pdf) (PDF). Connector 2000 Association. September 27, 2007. .

[7] "Notice of an Event of Default" (http://www.southernconnector.com/pdfs/12308 Event Notice- Trustee.pdf) (PDF). U.S. Bancorp. January 23, 2008. .

[8] "Event Notice No. 2010-1" (http://www.southernconnector.com/pdfs/Event Notice 2010-1.pdf) (PDF). Connector 2000 Association. January 11, 2010. .

[9] "Balance Sheet" (http://www.southernconnector.com/pdfs/Update - September 30 2009.pdf) (PDF). Connector 2000 Association. September 30, 2009. .

[10] Bathon, Michael; McCarty, Dawn (June 25, 2010). "Connector 2000 Association Files Bankruptcy in South Carolina" (http://www.businessweek.com/news/2010-06-25/connector-2000-association-files-bankruptcy-in-south-carolina.html). Businessweek. .

[11] "Palmetto Pass FAQ - Where can I use my Palmetto Pass?" (http://www.southernconnector.com/Zpalpass_faqs.htm). Southern Connector Toll Road. .

External links

- SouthernConnector.com (http://SouthernConnector.com)
- PalmettoPass.net (http://www.PalmettoPass.net)

South_Carolina_Department_of_Transportation

South Carolina Department of Transportation (SCDOT)	
Agency overview	
Formed	May 13, 1977
Preceding agencies	South Carolina Highway Commission South Carolina Department of Transportation and Highway Safety
Headquarters	Columbia
Agency executive	H.B. "Buck" Limehouse, Jr., Secretary of Transportation
Website	
[1]	

The **South Carolina Department of Transportation (SCDOT)** is a government agency in the US state of South Carolina. Its mission is to build and maintain roads and bridges and administer mass transit services.

By state law,[2] the SCDOT's function and purpose is the systematic planning, construction, maintenance, and operation of the state highway system and the development of a statewide mass transit system that is consistent with the needs and desires of the public. The SCDOT also coordinates all state and federal programs relating to highways. The goal of the SCDOT is to provide adequate, safe, and efficient transportation services for the movement of people and goods.

History

The South Carolina Department of Transportation is still familiarly known as the **Highway Department**, which is what the agency was called until May 13, 1977 when an act of the South Carolina General Assembly reformed the agency as the **Department of Highways and Public Transportation** (SCDHPT).[3]

The current name, the Department of Transportation, was established in the State Government Restructuring Act of 1993.[4] This act split functions of the SCDHPT to establish the SCDOT and the Department of Public Safety.

The roots of the state agency trace back to the establishment of a five member highway commission in 1917. Prior to 1917, county governments were entirely responsible for building and maintaining roads. The Federal Aid Road Act of 1916 and the promise of federal money to build highways served as the impetus for the creation of the commission. However, the commission lacked "the authority to designate roads to be improved with federal funds and the power to supervise directly the work being done."[5]

Organization

Commission and Secretary

The SCDOT is a department of the state government in the executive branch that reports to a cabinet secretary and a seven-member commission.[6] The Governor appoints one member of the commission and the **Secretary of Transportation**, who is a member of the Governor's cabinet.[2] The current secretary is H. B. "Buck" Limehouse, Jr.[7]

The other six commissioners each represent a district. The commission districts coincide with the congressional districts of South Carolina. The commissioner is appointed for a term of four years by legislative delegation for that district, that is, the members of the General Assembly that represent voters in that district. The chairman is selected from among all seven members by a vote of the commission.

The secretary hires the division heads, who are known as Deputy Secretaries.

Divisions of the SCDOT

The SCDOT has at least three divisions: Mass Transit; Construction, Engineering, and Planning; and Finance and Administration. The commission has the authority to establish additional divisions or to split these statutory divisions into multiples.

The SCDOT is a centralized government agency. Planning, design, procurement, finance and human resource functions all operate from the central office, or headquarters, in the state capitol of Columbia. The headquarters building is named for Silas N. Pearman, a former state highway engineer and chief commissioner of the agency.

Mass Transit Division

The Division of Mass Transit supports the development of a mass transit system and administers the state and federal aid mass transit programs.

Engineering Division

The Division of Construction, Engineering, and Planning comprises the largest part of the agency. The Deputy Secretary for this division is traditionally known as the **State Highway Engineer**. This position is generally recognized as the second-ranking person in the agency.

Finance and Administration Division

The Division of Finance and Administration is responsible for the federal aid reimbursements and other financial matters of the agency.

Engineering Districts

The SCDOT field offices are divided into seven districts headed by a **District Engineering Administrator**. The engineering district lines do not follow the same lines as the commission districts. Each District has responsibility for the maintenance, construction, traffic, and equipment (mechanical) operations within its boundaries. A district will oversee six to eight counties.

District offices are located in Columbia (District #1), Greenwood (#2), Greenville (#3), Chester, (#4), Florence (#5), Charleston (#6), Orangeburg (#7).

References

- Moore, John Hammond (1987), *The South Carolina Highway Department, 1917–1987*, Columbia, South Carolina: University of South Carolina Press, ISBN 0-87249-528-0

[1] http://www.scdot.org

[2] South Carolina Code of Laws, *Title 57 – Highways, Bridges and Ferries, Section 57-1-30* (http://www.scstatehouse.net/code/t57c001. htm),

[3] South Carolina General Assembly (May 13, 1977), *Act 82 of the 102nd Session* (http://www.scstatehouse.net/cgi-bin/web_legislation. exe), , retrieved April 10, 2007

[4] South Carolina General Assembly (June 18, 1993), *Act 181 of the 110th Session* (http://www.scstatehouse.net/cgi-bin/web_legislation. exe), , retrieved April 10, 2007

[5] Hammond, p. 47.

[6] South Carolina Code of Laws, *Title 1 – Administration of the Government, Section 1-30-10(B)(1)(ii)* (http://www.scstatehouse.net/code/ t01c030.htm),

[7] Inside SCDOT – Administration (http://www.scdot.org/inside/administration.shtml)

External links

- SCDOT Official Site (http://www.scdot.org)
- South Carolina State Government Site (http://www.sc.gov)

U.S._Route_29

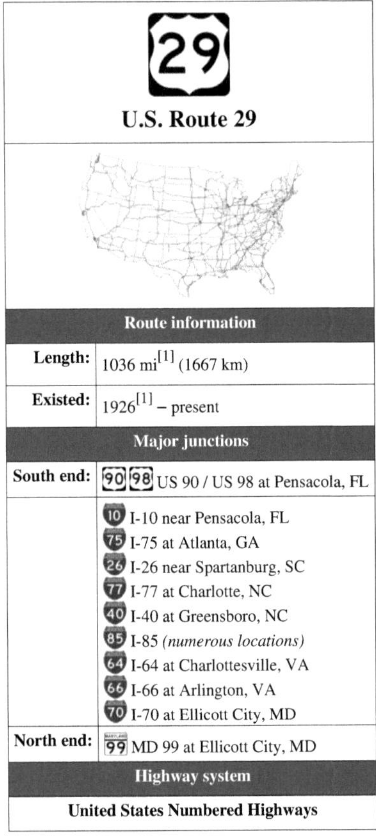

U.S. Route 29

Route information	
Length:	1036 mi[1] (1667 km)
Existed:	1926[1] – present
Major junctions	
South end:	US 90 / US 98 at Pensacola, FL
	I-10 near Pensacola, FL I-75 at Atlanta, GA I-26 near Spartanburg, SC I-77 at Charlotte, NC I-40 at Greensboro, NC I-85 *(numerous locations)* I-64 at Charlottesville, VA I-66 at Arlington, VA I-70 at Ellicott City, MD
North end:	MD 99 at Ellicott City, MD
Highway system	
United States Numbered Highways	

U.S. Route 29 (**US 29**) is a north–south United States highway that runs for 1036 miles (1667 km) from the western suburbs of Baltimore, Maryland, to Pensacola, Florida. This highway's northern terminus is at Maryland Route 99 in Ellicott City, Maryland. Its southern terminus is at US 90 and US 98 in Pensacola, Florida.

The section of US 29 between Greensboro, North Carolina, and Danville, Virginia, has been designated as Future Interstate 785 and has received "Future Interstate" signs in several locations along that route. It will become an official Interstate Highway once improvements have been completed.

From Greensboro, North Carolina to Tuskegee, Alabama, Interstate 85 (I-85) runs parallel with US 29, which along that stretch, serves primarily as a local route.

Route description

Lengths

	mi	km
FL	43	69
AL	226.550	365
GA	207	333
SC	106.4	171
NC	168	270
VA	248	399
DC	55	89
MD	25.859	42
Total	1,036	1,667

Florida

US 29 begins at U.S. Route 90 and U.S. Route 98 in downtown Pensacola, Florida. Throughout the state, U.S. 29 is twinned with the unsigned State Road 95.

The entire route in Florida runs within Escambia County. From its terminus north to State Road 296, it is known as North Palafox Street. From this point it is known as Pensacola Boulevard north to Ten Mile Road, approximately one mile north of U.S. 90 Alternate. Between SR 296 and the Molino community, U.S. 29 runs parallel to its former routing, which is now Escambia County Road 95A (Old Palafox Highway).

Alabama

US 29, internally designated by the Alabama Department of Transportation as **State Route 15** (SR-15), is a southwest-northeast state highway across the southeastern part of the U.S. state of Alabama. SR-15 ends in Brewton at a junction with US-31 (SR-3) and SR-41, but US-29 continues west with US-31/SR-3 to Flomaton and south on SR-113 to the Florida state line.

U.S. Highway 29 and SR-15 traverse Alabama in a general northeast/southwest slope. It has never been a major route in the state; its significance was completely overshadowed with the completion of Interstate 65 and Interstate 85 during the 1970s. Today, US-29 and SR-15 serve primarily to connect numerous smaller towns and cities in the southwest, south-central, and eastern parts of Alabama.

Georgia

US 29 passes through the northern portion of Georgia, serving Atlanta and Athens. The highway passes by notable universities, such as Georgia Tech in Atlanta and the University of Georgia in Athens. US 29 also meanders through Hartwell and the Lake Hartwell region near the South Carolina border. From West Point, Georgia (Just south of LaGrange, Georgia) at the Alabama-Georgia Line to downtown Atlanta, Georgia State Route 8 and Georgia State Route 14 are paired with US 29 at various points in the state. US 29 has also been named Roosevelt Hwy.

South Carolina

In South Carolina, US 29 maintains a northeasterly routing, passing through Anderson, Greenville, and Spartanburg.

From Greenville through Greer, US 29 is known as **Wade Hampton Boulevard**. It is a major commercial artery for both Greer and Taylors. A six-lane highway, the road forms the western border of Bob Jones University and then passes near Chick Springs, a mineral springs that served as the focus of a small but important resort community during the nineteenth century.

US 29 was built as the main highway between Greenville and the other city of northwestern South Carolina, Spartanburg. The construction of Interstate 85 connecting Greenville to Spartanburg left US 29 underused until recent decades.

North Carolina

In North Carolina, US 29 connects the cities of Charlotte, Concord, and Greensboro. US 29 routes through Charlotte along Tryon Street, one of the main arteries that runs through uptown Charlotte.

Virginia

In Virginia, part of U.S. 29 is named the Lee Highway. U.S. 29 connects the historic small cities and large towns of west-central Virginia, including Danville, Lynchburg, Charlottesville, Culpeper, Warrenton, Manassas, and Fairfax, with Arlington, Virginia, and Washington, D.C., to the northeast, and with North Carolina to the southwest.

Along its route in Virginia, U.S. 29 provides significant access to and from several major colleges and universities, including the University of Virginia in Charlottesville, George Mason University in Fairfax, Sweet Briar College in Sweet Briar, and Liberty University, Lynchburg College, and Randolph College in Lynchburg.

District of Columbia

US 29 enters the District of Columbia and Washington, D.C., via the Francis Scott Key Bridge adjacent to Georgetown University. The designation turns east onto the Whitehurst Freeway, bypassing Georgetown to the south. Upon crossing Rock Creek, the freeway ends, becoming the at-grade K Street. US 29 remains on K Street to 11th Street, where US 29 turns north onto 11th for seven blocks. At Rhode Island Avenue, US 29 turns right. US 29 northbound turns left at 6th Street NW (touching US 1 where it turns from Rhode Island Avenue to 6th Street); it follows 6th Street NW for two blocks and then turns left onto Florida Avenue NW, where it then turns right onto Georgia Avenue NW. US 29 southbound at this point, however, follows 7th Street, NW to Rhode Island Avenue NW. The route maintains a northerly routing as it passes through northern Washington, D.C. and enters Maryland. During its alignment with Georgia Avenue NW, US 29 bypasses the Howard University campus to the west.

Maryland

In Maryland, US 29 turns northeast onto Colesville Road, interchanges with the Capital Beltway (Interstate 495), becomes Columbia Pike, and interchanges with New Hampshire Avenue (Maryland Route 650), Maryland Route 198, Maryland Route 32, Maryland Route 175, Maryland Route 100, US 40 and I-70 before terminating at Maryland Route 99 northwest of Ellicott City. There are plans to extend the northern terminus to Taneytown, near the Pennsylvania-Maryland border.

History

Warrenton Turnpike is the former name of US 29 through Prince William County, Virginia. This is the name that was used for this road during the Civil War. Although the road has been expanded past Manassas into four lanes, it remains a rural two lane highway through Manassas National Battlefield Park, where Interstate 66 carries through traffic. On either side of the road through the battlefield, split rail fences define property borders.

A US 29 shield
used in Florida
prior to 1993

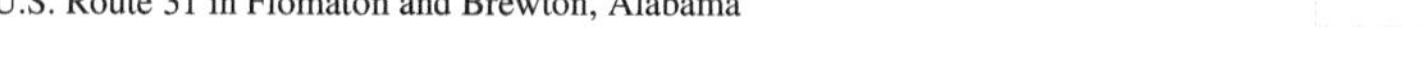

U.S. Route 170

The portion of US 29 from US 70 at Charlotte, North Carolina northeast to Lynchburg, Virginia was **U.S. Route 170** from 1926 until 1931, when US 29 was extended over it.[2]

US 170 (1926-1931)

Major intersections

- Interstate 10 in Ensley, Florida
- U.S. Route 31 in Flomaton and Brewton, Alabama

- U.S. Route 84 in Andalusia, Alabama
- U.S. Route 82 in Union Springs, Alabama
- U.S. Route 80 in Tuskegee and near Auburn, Alabama
- U.S. Route 27 in Lagrange, Georgia
- U.S. Route 27 Alternate in Newnan, Georgia
- Interstate 75/Interstate 85 (the Downtown Connector) in Atlanta, Georgia
- U.S. Route 78 at Decatur, Georgia
- Interstate 285 at Tucker, Georgia
- U.S. Route 78/U.S. Route 129/U.S. Route 441 at Athens, Georgia
- State Route 17 (Georgia) at Royston, Georgia
- State Route 77 (Georgia) at Hartwell, Georgia
- U.S. Route 76/U.S. Route 178 at Anderson, South Carolina
- U.S. Route 25/U.S. Route 123/U.S. Route 276 at Greenville, South Carolina
- Interstate 26 at Spartanburg
- U.S. Route 74 in Gastonia and Charlotte, North Carolina
- Interstate 77/U.S. Route 21 in Charlotte, North Carolina
- Interstate 485 in Charlotte, North Carolina
- Interstate 85 in Concord, North Carolina
- U.S. Route 52 in Salisbury and Lexington, North Carolina
- Interstate 85 in Greensboro, North Carolina
- U.S. Route 70 in Greensboro and Salisbury, North Carolina
- Interstate 40 in Greensboro, North Carolina
- U.S. Route 58/U.S. Route 360 in Danville, Virginia
- U.S. Route 60 in Amherst, Virginia
- Interstate 64 in Charlottesville, Virginia
- U.S. Route 250 in Charlottesville, Virginia

- U.S. Route 33 in Ruckersville, Virginia
- U.S. Route 15 in Culpeper and Gainesville, Virginia
- U.S. Route 17 in Opal and Warrenton, Virginia
- Interstate 66 at Gainesville, Virginia
- U.S. Route 50 at Fairfax, Virginia
- Interstate 495 at Silver Spring, Maryland
- Interstate 70 at Ellicott City, Maryland

See also

- U.S. Route 129

References

[1] US Highways from US 1 to US 830 (http://www.us-highways.com/us1830.htm) Robert V. Droz

[2] Federal Highway Administration, U.S. 29 Maryland to Florida (http://www.fhwa.dot.gov/infrastructure/us29.htm)

	1	2	3	4	5	6	7	8	9	10	11	12	13	14	15	16	17	18	19
												Main U.S. Routes							
20	21	22	23	24	25	26	27	28	29	30	31	32	33	34	35	36	37	38	
40	41	42	43	44	45	46		48	49	50	51	52	53	54	55	56	57	58	59
60	61	62	63	64	65	66	67	68	69	70	71	72	73	74	75	76	77	78	79
80	81	82	83	84	85		87		89	90	91	92	93	94	95	96	97	98	99
101		163						400				412				425			
Lists		U.S. Routes • Bannered • Divided • Bypassed																	

Browse numbered routes

Interstate_385

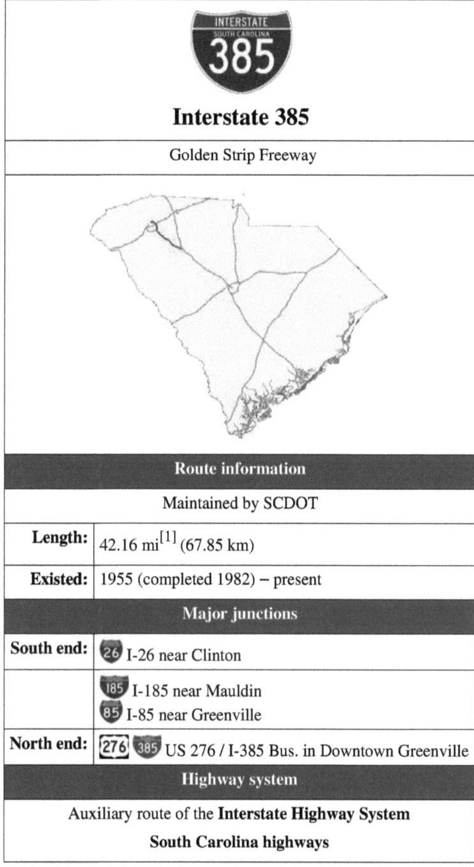

Interstate 385	
Golden Strip Freeway	
Route information	
Maintained by SCDOT	
Length:	42.16 mi[1] (67.85 km)
Existed:	1955 (completed 1982) – present
Major junctions	
South end:	I-26 near Clinton
	I-185 near Mauldin I-85 near Greenville
North end:	US 276 / I-385 Bus. in Downtown Greenville
Highway system	
Auxiliary route of the **Interstate Highway System** **South Carolina highways**	

Interstate 385 (I-385) is an Interstate Highway located in The Upstate region of South Carolina. It goes from Clinton, South Carolina at Interstate 26 to Greenville, South Carolina at Laurens Road/U.S. 276 (Exit 42). After exit 42, Interstate 385 turns into a Business Spur and becomes East North Street and later — for northbound motorists only — Beattie Place. The spur promptly ends at U.S. 29 (Church Street) near the Bi-Lo Center in downtown Greenville.

The explosive economic growth of southern Greenville county is largely attributed to I-385 and its connection to the city of Greenville and the major cities of Atlanta and Charlotte (via I-85). This area is known by locals as the "Golden Strip".

Route description

Highway 49 exit on I-385, four miles before the road ends at Interstate 26.

I-385 features a rather unusual rest area in the median strip near Laurens, that serves both directions of traffic. It was completed as part of the original design of the U.S. 276 expressway in 1958, modeled after the type of single median-located rest areas shared by both north and southbound traffic (to save money). The design is similar to many of those built on turnpikes that predated the Interstate System.

Business Route

Interstate 385 Business

Location:	Greenville, South Carolina

Interstate 385 starts where it meets the Interstate 26, near Clinton and heads in a northwest direction where it ends with exit 42, near downtown Greenville. After exit 42, I-385 turns into **Business Spur 385** that then promptly ends at U.S. 29 near the Bi-Lo Center in downtown Greenville. It is one of five official Interstate business spurs in S.C. (the others being spurs of I-20, I-126, I-526, and I-585).

History

The general idea — but none of the specifics — of I-385 were present on the 1955 Yellow Book map of the Greenville area. Also of note is that Interstate 85 would have used the U.S. Route 29 corridor from Greenville east towards Spartanburg based on the diagram.

The portion of I-385 that replaced U.S. 276 (from SC-417 in Mauldin to SC-56 / I-26 in Clinton) was initially the first phase built of an SC DOT plan that predated the Interstate System to upgrade and bypass existing through routes, the goal of forming a single limited-access highway from Greenville to the port of Charleston via the State Capital of Columbia. This plan was scrapped as soon as the future I-26 was added to the act of Congress that set into motion the Interstate System. As a result, I-26 was one of the first Interstates in the south to open in significant mileage (most in SC between 1959 and 1963).

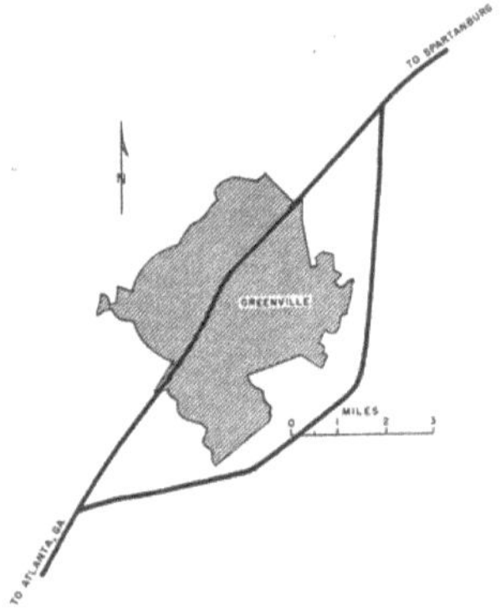

1955 "Yellow Book" map of Greenville

Prior to 1985, I-385 was only signed as such from downtown Greenville to I-85. The portion of the freeway from U.S. 276 in Mauldin to the southern terminus at I-26 was signed as U.S. 276. When the connecting portion was completed, the entire freeway was signed as I-385.

For seven months ending July 23, 2010, northbound traffic could not use a 15-mile section of I-385 in Laurens County due to a $60.9 million project to pave the portion extending from highway 101 to the I-385-I-26 interchange near Clinton, SC in concrete. The closing of a major highway generated controversy.[2]

Exit list

County	Location	#	Destinations	Notes
Laurens	Clinton	1	I-26 east – Columbia	Southern terminus
		2	SC 308 to I-26 west – Ora, Clinton	Northbound signed SC 308 only
	Laurens	5	SC 49 – Laurens, Cross Anchor	
		9	US 221 – Laurens, Enoree	
	Gray Court	10	Metric Road – Gray Court	
		16	SC 101 – Woodruff, Gray Court	
		19	SC 14 east – Owings, Gray Court	South end of SC 14 overlap
	Fountain Inn	22	SC 14 west – Fountain Inn	North end of SC 14 overlap
Greenville		23	SC 418 – Fountain Inn, Pelzer	
		24	Fairview Street – Fountain Inn	
		26	Harrison Bridge Road	Southbound exit and entrance
	Simpsonville	27	Fairview Road – Simpsonville	
		29	Georgia Road – Simpsonville	
	Mauldin	30	I-185 north / US 276 north – Mauldin, Anderson	
		31	SC 417 (Laurens Road) – Mauldin	No southbound exit; southbound exit is via exit 30
		33	Bridges Road – Mauldin	
		34	Butler Road – Mauldin	
		35	SC 146 (Woodruff Rd)	
	Greenville	36	I-85 – Spartanburg, Anderson	Signed as exits 36A (north) and 36B (south) northbound Single exit 36 southbound
		37	Roper Mountain Road	
		39	Haywood Road	
		40	SC 291 (Pleasantburg Dr.)	Signed as exits 40A (south) and 40B (north)
		42	US 276 – Travelers Rest, Mauldin	
		44	Bryce Avenue, North Street	Signed as exits 44A (east/north) and 44B (west/south) northbound
			North Street north	Northern terminus

References

[1] Route Log - Auxiliary Routes of the Eisenhower National System Of Interstate and Defense Highways - Table 2 (http://www.fhwa.dot. gov/reports/routefinder/table2.htm)

[2] Cay, Nathaniel (2010-07-25). "I-385 reopened after 7-month closure" (http://www.thesunnews.com/2010/07/25/1601956/ i-385-reopened-after-7-month-closure.html). *Greenville News*. . Retrieved 2010-07-25.

Interstate_85

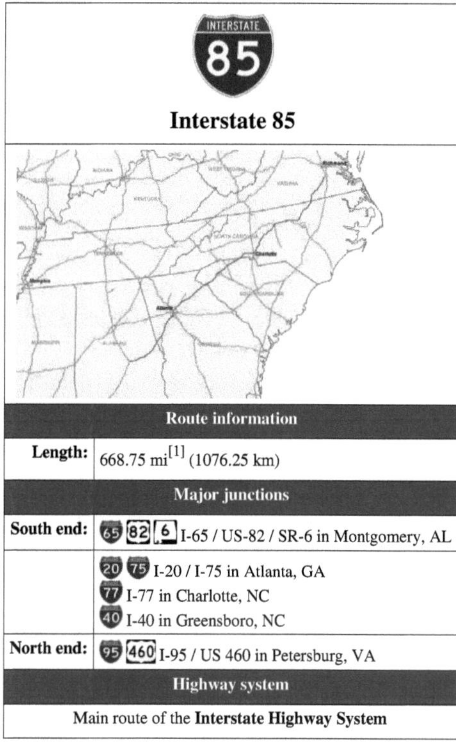

Interstate 85 (I-85) is a major interstate highway in the Southeastern United States. Its current southern terminus is at an interchange with Interstate 65 in Montgomery, Alabama; its northern terminus interchanges with Interstate 95 in Petersburg, Virginia, near Richmond. In its northern half it roughly parallels an ancient indian trading path documented since colonial times from Petersburg, Virginia to the Catawba Indian territory.

An extension of Interstate 85 is proposed west from Montgomery to interchange with Interstate 20 & Interstate 59 just east of the Mississippi state line.[2]

Route description

Lengths

	mi[1]	km
AL	80	130
GA	179.9	292
SC	106.28	172
NC	233.93	377
VA	68.64	112
Total	668.75	1,082

Alabama

Interstate 85 begins as a fork off Interstate 65 in Montgomery. From here, I-85 parallels U.S. Route 80 until the highway nears Tuskegee. At Tuskegee, I-85 leaves U.S. 80 and starts to parallel U.S. Route 29, which the highway parallels for much of its length.

I-85 also passes near Auburn, Opelika, Valley and Lanett before crossing the Chattahoochee River into Georgia.

Georgia

Alabama state line to LaGrange

In Georgia, Interstate 85 bypasses West Point before coming into the LaGrange area. East of LaGrange, I-85 intersects Interstate 185. Travelers can take I-185 to Columbus and Fort Benning. Travelers can also see the construction of Kia's multi-billion dollar plant in West Point, just a few miles south west of LaGrange.

I-85 in Georgia is designated — but not signed as - State Route 403.

Interstate 85 under partial construction near the Alabama state line.

LaGrange to Atlanta

From LaGrange, I-85 heads northeastward towards Atlanta. Before reaching Atlanta, the highway drives through a recently widened stretch between exits 28 and 74, passing the suburbs of Newnan, College Park and East Point as well as intersecting Interstate 285 and providing access to Hartsfield-Jackson Atlanta International Airport. On the south side of the city, I-85 merges with Interstate 75 to form the Downtown Connector. Just before downtown, the highway goes through a pile-up intersection with I-20. Afterward, the two highways go through the heart of the city until they split and cross over each other just north of Georgia Tech. At the split, I-75 exits via the right three lanes and then heads northwest while I-85 uses the left three lanes and then heads northeast.

Atlanta to South Carolina state line

Continuing northeast of Atlanta, Interstate 85 continues through the northeastern suburbs, bypassing Chamblee and Doraville, where there is another intersection with I-285 (nicknamed Spaghetti Junction). Also, on the section from Chamblee-Tucker Road to Old Peachtree road in Gwinnett County, the current HOV lanes presented are being considered for converting to high occupancy toll lanes to help ease the congestion which is currently present during rush hours.

I-75 co-signed with I-85 in downtown Atlanta

Continuing northbound, drivers will see another interstate spur - Interstate 985, which provides a link to Gainesville - before heading through northeastern Georgia. At Lake Hartwell - which was formed by the damming of the Savannah River - I-85 crosses into South Carolina.

Just northeast of the I-75/I-85 northern split, I-85 was rerouted just north of its original route in the 1990s from just northeast of the interchange to near the State Route 400 interchange.[3] The original route is now signed State Route 13.

South Carolina

Interstate 85 provides the major transportation route for the Upstate of South Carolina, linking together the major centers of Greenville and Spartanburg with regional centers of importance.

Georgia state line to Greenville

In South Carolina, Interstate 85 bypasses Clemson and Anderson on the way to Greenville. Beginning at Anderson, I-85 widens from four to six lanes. Near Powdersville, U.S. 29 joins I-85 and they run concurrently until they cross the Saluda River.

Interstate 85 North after Exit #1 in Oconee County, South Carolina in 2008.

Greenville to Spartanburg

Interstate 85 bypasses just south of Greenville, but provides two links into the city via spur routes Interstates 185 and 385. U.S. 29 splits from I-85 and joins I-185 toward downtown Greenville. I-185 recently saw a major expansion as a toll road south of the city as the Southern Connector was completed. In this section, the roadway is six to eight lanes wide, depending on the location.

Much of the Upstate's industrial, institutional, and commercial base is located along I-85, especially with regards to the growing automotive industry. The North American headquarters for Michelin, along with BMW's Spartanburg assembly plant, and the new Clemson University International Center for Automobile Research are all visible from the road. Additional companies have major operations along this corridor, including Hubbell Lighting, Sara Lee Corporation, Hitachi, and General Electric. I-85 also bypasses Greenville-Spartanburg International Airport, which serves the Greenville-Spartanburg metropolitan area.

Spartanburg to North Carolina state line

Instead of going into Spartanburg, I-85 bypasses the city to the north. Its original route is now signed Business Loop 85 and was approved by AASHTO on April 22, 1995.[4]

North of Spartanburg, I-85 narrows from six lanes back to four lanes and bypasses Gaffney. At Gaffney, motorists can see The Peachoid, a large water tower with its top shaped like a peach, which is one of the state's most important crops. Just northeast of Exit 95 in the median is a cemetery[5]

Much of the terrain between Spartanburg and the North Carolina border is rural in nature.

Interstate 85 southbound as it enters South Carolina near Blacksburg.

North Carolina

In North Carolina, I-85 merges with I-40 from Greensboro to Hillsborough, just west of Durham. In Alamance County, the highway is also known as the Sam Hunt Freeway.

South Carolina state line to Charlotte

Immediately upon coming into North Carolina, drivers will realize that Charlotte, North Carolina is still 45 minutes away. This is because Interstate 85 takes a more west to east routing along with U.S. Route 29 and U.S. Route 74, nears Kings Mountain.

Drivers will go through Gastonia and will eventually reach Charlotte with the option to bypass using Interstate 485.

Charlotte to Greensboro

In Charlotte, I-85 bypasses Charlotte-Douglas International Airport and turns northeastward just before reaching uptown Charlotte; thus I-85 just bypasses uptown to the north. The part of this section from near Little Rock Road northward to the US-29 / NC-49 split became the first mileage of the entire I-85 corridor to be completed (1958).

The junction between I-85 and Interstate 77 north of uptown Charlotte is a strange configuration. While I-85 passes under I-77, the northbound lanes of I-77 are to the west of I-77's southbound lanes. The travel lanes on I-77 return to their proper positions north and south of this interchange.

North of Charlotte, the highway passes near Concord, where one can see the Charlotte Motor Speedway. There is much dedication to the Earnhardt family as there are many roads named using his name, such as Dale Earnhardt Boulevard and North Carolina Highway 3.

At milepost 98, the northbound lanes of I-85 cross under the southbound lanes, and cross back to the correct configuration near milepost 102. This results in motorists driving to the left of opposing traffic for approximately three miles (5 km). The switch is not very noticeable, because the roadways are separated by up to 400 feet (123 m) of woods in this area. A rest area and Vietnam Veterans memorial are located in the median of this section, so the crossover allows for all exits into the rest area to be normal right-hand exits.

Greensboro to Durham

Continuing northbound, I-85 passes though or near Salisbury, Lexington and High Point before reaching Greensboro. At Greensboro, I-85 shifts to its to a new routing, away from downtown. At Exit 120, Business I-85 (old I-85 through town) exits and Interstate 73 begins its short overlap with I-85, leaving at Exit 122 (although the two interstates never share the same alignment). I-85 intersects with a new I-73 freeway for travelers bound westward towards Winston-Salem (via I-40). The former I-85 routing nearest downtown Greensboro (now Business I-85/I-40 (Business 85 joins I-40 at "Death Valley")) is also notorious for many traffic accidents and earned the above mentioned

Interstates 85/40 through Burlington, NC.

nickname "Death Valley". (Until September 2008, I-40 departed from I-85 at Exit 120, and its former — and once again current — alignment was signed as Business 40.)

At Exit 131, I-85 joins I-40 east of downtown, and the two highways are cosigned as they pass through Burlington, Graham and Mebane then separate near Hillsborough. Interstate 85 continues to Durham while Interstate 40 (Exit 163) turns toward Chapel Hill, Cary and Raleigh.

Durham to Virginia state line

From Durham, I-85 turns northeastward and heads toward Virginia. Until replaced during a recent upgrade, signs for Richmond used to exist in this part of North Carolina even though the end of the road is in Petersburg, Virginia.

Virginia

Starting from the Virginia border, drivers will pass South Hill and McKenney before heading into a large forest of trees. After the forest, Interstate 85 reaches Petersburg and ends at Interstate 95. The highway is briefly cosigned with U.S. Route 460 from a few miles west of Petersburg in Dinwiddie County to I-95.

The last four miles (6 km) of I-85 near Petersburg once formed the southern end of the Richmond-Petersburg Turnpike, which was completed in 1958. The tolls were removed in 1992 after Interstate 295 was completed.[6]

Before a 2010 decision to raise the speed limit in the state to 70, Virginia's portion of I-85 was also the only Interstate Highway in the state with a posted speed limit greater than 65 miles per hour (105 kilometers per hour). It was raised from 65 mph (105 km/h) to 70 mph (113 km/h) on July 1, 2006, by the state legislature.

The Northern Terminus of I-85 is at the main interstate highway of the east coast, I-95, which is a gateway to various major cities, including Miami, Jacksonville, Washington DC, Philadelphia, New York City, Boston, etc.

History

I-85 was rerouted around Greensboro in 2004; and it split with I-40 eight miles (13 km) east of the original departure point. I-40 ran with I-85 along the bypass to the southern/western end and I-40 continued on a new freeway alignment at Exit 121 until September 2008, when it was rerouted back to its old alignment through the city. Despite its reroute around Greensboro, the overall length for I-85 in North Carolina (233 miles/373 km) remains the same as before.

Future

There is currently a plan by Alabama state transportation officials to extend I-85 across western Alabama, where it will connect with I-20 and I-59 near Cuba, Alabama. This extension will roughly follow the route of U.S. 80, going through or bypassing Selma and Demopolis.[7] The FHWA approved the alignment on February 17, 2011 after AASHTO approved at its Fall 2010 meeting in Biloxi, Mississippi. Also approved was the proposal to re-designate part of existing I-85 south and east of Montgomery to be bypassed as part of the extension of I-85 as I-685. Alabama has permission to co-sign this part of I-85 as I-685 until the new alignment is built. [8] This section is also envisioned by some as part of a proposed Interstate 14.

Interstate 85 is scheduled to have several new auxiliary routes in the future. Interstate 285 is also planned to follow part of the U.S. Route 52 freeway from Lexington to Winston-Salem, both in North Carolina. Interstate 785 is currently planned by the North Carolina Department of Transportation to run from Greensboro to Danville, Virginia. The proposed route would follow the current U.S. 29 corridor. There are plans for I-85 from Anderson County, South Carolina to Spartanburg County, South Carolina to become four to five lanes in each direction including HOV lanes, if it is funded it will start construction in 2012.

Major intersections

- Interstate 65 in Montgomery, Alabama
- Interstate 185 near LaGrange, Georgia.
- Interstate 285 near College Park, Georgia. I-85 intersects this highway twice.
- Interstate 75 in Atlanta. The two highways run concurrently through much of the city.
- Interstate 20 in Atlanta.
- Interstate 985 near Lawrenceville, Georgia.
- Interstate 185 near Greenville, South Carolina.
- Interstate 385 near Greenville.
- Interstate 26 near Spartanburg, South Carolina.
- Interstate 585 near Spartanburg via Business Loop 85. An extension is currently underway that will extend I-585 to I-85.
- Interstate 485 in Charlotte, North Carolina. I-85 intersects I-485 twice.
- Interstate 77 in Charlotte, North Carolina.
- Interstate 40 in Greensboro, North Carolina. They stay connected until Hillsborough, North Carolina.
- Future Interstate 73 in Greensboro.
- Interstate 95 in Petersburg, Virginia. (Map [9])

See also

- Death Valley (North Carolina)
- Richmond-Petersburg Turnpike
- I-85 Corridor
- I-85 Rivalry

Business routes

- Interstate 85 Business Loop in High Point, North Carolina and Greensboro, North Carolina.
- Interstate 85 Business Loop near Spartanburg, South Carolina.

Pop culture

- Interstate 85 is mentioned by Bubba Sparxxx in a song by Petey Pablo called "Get on Dis Motorcycle". In the song Bubba recounts how Petey made him drive drunk all the way from LaGrange, GA, through Spartanburg, SC, and into High Point, NC.
- The Interstate Highway is referenced in John Mayer's song "Why Georgia", where the lyrics are "I am driving up 85 in the kind of morning that lasts all afternoon".

References

[1] "Route Log and Finder List — Interstate System: Table 1" (http://www.fhwa.dot.gov/reports/routefinder/table1.cfm). FHWA. . Retrieved 2007-09-26.

[2] Volkert and Associates, I-85 Extension Corridor Study & EUIS (http://www.i85extension.com/)

[3] State Route 13 Page (http://web.archive.org/web/20070523175311/http://www.geocities.com/garoadwarrior76/SR13_Profile.html) Peach State Roads. Retrieved 27 May 2007.

[4] Interstate 85@Interstate-Guide.com (http://www.interstate-guide.com/i-085.html) Courtesy AARoads. Retrieved 27 May 2007.

[5] Graveyard in the I-85 Median (http://www.roadsideamerica.com/tip/2837) RoadsideAmerica.com. Retrieved 27 December 2009.

[6] Kozel, Scott Richmond-Petersburg Turnpike (I-95/I-85) and I-285 (http://www.roadstothefuture.com/RPT_I295.html) Retrieved 27 May 2007.

[7] Hinnen, Jerry. Shelby shares views with Hale, Greene counties (http://www.demopolistimes.com/articles/2005/01/18/news/news02.txt) Posted by the Demopolis Times, 17 January 2005.

[8] FHWA letter downloaded from http://cms.transportation.org/sites/route/docs/Alabama%20Interstate%20FHWA%20Decision%20Letter. pdf April 14, 2011

[9] http://maps.google.com/maps?q=Petersburg,+Virginia&ll=37.241262,-77.449493&spn=0.492298,0.720154&hl=en

External links

- I-85 Extension Corridor Study (http://www.i85extension.com/) - Corridor study and environmental impact statement by the Alabama Department of Transportation and the Federal Highway Administration.

Browse numbered routes

U.S._Route_278

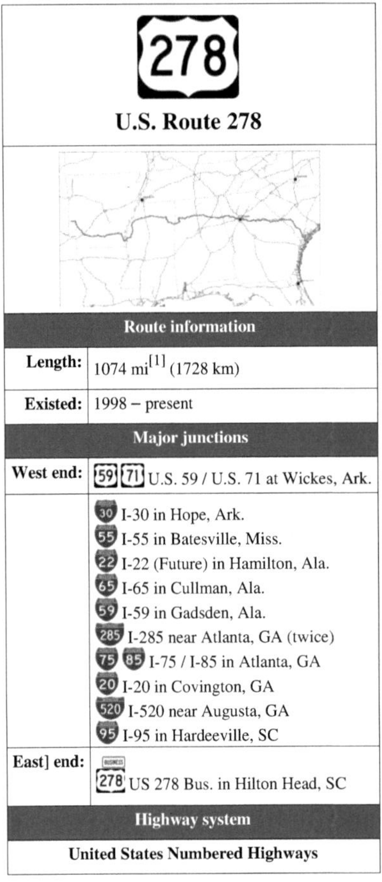

U.S. Route 278 is a parallel route of U.S. Route 78. It currently runs for 1,074 miles (1,728km) from Hilton Head Island, South Carolina to Wickes, Arkansas at U.S. Highway 71/U.S. Highway 59. It might be notable that it is longer than its parent highway, US Hwy-78. US Hwy-278 passes through the states of South Carolina, Georgia, Alabama, Mississippi, and Arkansas. This highway passes through the cities and towns of Augusta, Covington, Atlanta, Powder Springs, Georgia; Hiram, Georgia; Dallas, Georgia; Rockmart, Georgia; Cedartown, Georgia; Gadsden, Cullman, Alabama; Tupelo, Oxford, Greenville, Mississippi; Monticello, and Hope, Arkansas.

It also passes through the Savannah River Site, of the Department of Energy in South Carolina, and it crossed the Tennessee-Tombigbee Waterway in eastern Mississippi.

By coincidence, U.S. 278 passes through the places where many colleges, universities, and technical institutes are located, including the Georgia Institute of Technology, Georgia State University, and many others in Atlanta, Georgia; the Augusta State University; the University of Mississippi, and the University of Arkansas at Monticello.

Route description

Arkansas

US 278 begins at an intersection with U.S. Route 59/U.S. Route 71 in the town of Wickes in southwestern Arkansas. From Wickes, US 278 continues eastward through Hope, Camden, and Monticello to McGehee, where it meets U.S. Route 65. US 278 overlaps US 65 southward for 16 miles (26 km) until they separate in Lake Village. There, US 65 splits off and US 278 overlaps U.S. Route 82 east to the Mississippi River, where US 82 and US 278 cross into Mississippi.

Mississippi

US 82 and US 278 go through Greenville to Leland, where US 278 separates from US 82 at U.S. Route 61. US 278 then joins US 61 northward through Cleveland before splitting in Clarksdale. East of Clarksdale, it overlaps Mississippi Highway 6 through Batesville, Oxford, and Pontotoc before reaching Tupelo. At Tupelo, MS 6 separates from US 278 while US 278 overlaps U.S. Route 45 south to New Wren. From New Wren, US 278 continues east through Amory before entering Alabama.

Alabama

U.S. 278 enters Alabama between Greenwood Springs, Mississippi and Sulligent, Alabama. As in Georgia, this and all U.S routes are partnered with state routes; however, there are few instances throughout the state where the state route number is posted. From the Mississippi state line to Guin, U.S. 278 is paired with State Route 118. From Hamilton to the Georgia state line, U.S. 278 is paired with State Route 74.

U.S. 278 junctions U.S. Route 431 at Gadsden. The two routes overlap until they split at Attalla. After a reconfiguration of 3rd St SW at Main Ave SW to connect directly with 4th St SW in Cullman, U.S. 278 no longer briefly overlaps U.S. Route 31 for a block. U.S. 278 overlaps U.S. Route 43 between Hamilton and Guin. Prior to the completion of Corridor X (Future I-22), these two routes also overlapped U.S. Route 78 between these two towns, with East U.S. 78 travelling in the same direction as West U.S. 278.

Georgia

U.S. Route 278 (and S.R. 12) in downtown Crawfordville, GA.

In Atlanta, it (U.S. 278) runs along Donald Lee Hollowell Parkway (Bankhead Highway), North Avenue, briefly along Piedmont Road and Ponce de Leon Avenue. Outside the Perimeter, U.S. 278 runs along Covington Highway. In Lithonia, Georgia, at the intersection with Turner Hill Road (S.R. 124), U.S. 278 merges with Interstate 20 for 15 miles (24 km). It leaves I-20 at exit 90, in Covington, Georgia.

U.S. 278 is co-signed with a state route for its entire length in Georgia, as are all U.S. highways in Georgia. From the Alabama line into Lithia Springs, Georgia, it is co-signed with S.R. 6. From Lithia Springs through Atlanta, it is merged with U.S. Route 78/S.R. 8. In Atlanta, the federal highway merges further with other highways including S.R. 10 and U.S. Route 23 before splitting off at Ponce de Leon Ave. and East Lake Road in Decatur. In Avondale Estates, Georgia, S.R. 12 is paired with U.S. 278, all the way into Thomson, Georgia. There, the federal route merges with U.S. 78/S.R. 10 to the South Carolina line.

South Carolina

After crossing the Savannah River, U.S. 278 bypasses North Augusta, South Carolina en route to Beech Island and Johnson Crossroads. It then winds through a corner of the Savannah River Site. The route then continues onward through the communities of Allendale, Fairfax, Hampton, Varnville, Ridgeland, and Hardeeville where it meets I-95. U.S. 278 shares the route between Ridgeland and Hardeeville with U.S. 17. Prior to the construction of Exit 8 on Interstate 95 (which provides direct access to southern Beaufort County), U.S. 278 went from Ridgeland through Old House and Okatie toward Bluffton, along present-day state routes 336, 462, and 170.

Upon reaching Hardeeville, the route heads eastward toward the Atlantic with major new developments lining the spine of the road from Hardeeville through Sun City and Bluffton before crossing over the bridge to Hilton Head Island. U.S. 278 ends at U.S. 278 Business on the southern portion of Hilton Head Island, just outside Sea Pines Plantation. Old U.S. 278 was routed along what is now U.S. 278 Business until 1998, when a new toll road, dubbed the **Cross Island Parkway** was built. U.S. 278 was routed along the new parkway. The completion of the "Gateway to Hilton Head", a direct expressway connection from the north side to the south side of the island, has caused a great amount of commercial and residential development along the road.

History

Until early 2005, U.S. 278 was only signed in Mississippi from the Alabama state line to U.S. Route 45 west of Amory, Mississippi. US 278 in western Alabama used to be routed through Haleyville by what are now Alabama highways 195 and 129.

Major intersections

- U.S. Route 59/U.S. Route 71 at Wickes, Arkansas
- U.S. Route 70 at Dierks, Arkansas
- U.S. Route 371 at Nashville, Arkansas
- Interstate 30 at Hope, Arkansas
- U.S. Route 67 at Hope, Arkansas
- U.S. Route 371 at Rosston, Arkansas
- U.S. Route 79 at Camden, Arkansas
- U.S. Route 167 at Hampton, Arkansas
- U.S. Route 63 at Warren, Arkansas
- U.S. Route 425 at Monticello, Arkansas
- U.S. Route 65 from McGehee to Lake Village, Arkansas
- U.S. Route 82 from Lake Village, Arkansas to Leland, Mississippi
- U.S. Route 61 from Leland to Clarksdale, Mississippi
- U.S. Route 49 at Clarksdale, Mississippi
- Interstate 55 at Batesville, Mississippi
- U.S. Route 45 at Okolona, Mississippi
- U.S. Route 78 (future Interstate 22) at Hamilton, Alabama
- U.S. Route 43 from Guin to Hamilton, Alabama
- Interstate 65 at Cullman, Alabama
- U.S. Route 31 at Cullman, Alabama
- U.S. Route 231 near Snead, Alabama
- U.S. Route 11 at Attalla, Alabama
- U.S. Route 431 from Attalla to Gadsden, Alabama
- Interstate 59 at Gadsden, Alabama
- U.S. Route 27 at Cedartown, Georgia

- U.S. Route 78 from Lithia Springs to Druid Hills, Georgia
- Interstate 75/Interstate 85 at Atlanta, Georgia
- Interstate 285 at Decatur, Georgia
- U.S. Route 129/U.S. Route 441 at Madison, Georgia
- Interstate 20 at Barnett, Georgia
- U.S. Route 221 at Harlem, Georgia
- U.S. Route 25 at Augusta, Georgia
- Interstate 520 east of Augusta, Georgia
- U.S. Route 301 at Allendale, South Carolina
- U.S. Route 321 at Fairfax, South Carolina
- U.S. Route 601 at Hampton, South Carolina
- U.S. Route 17 from Ridgeland to Hardeeville, South Carolina
- Interstate 95 at Hardeeville, South Carolina

See also

- U.S. Route 78

References

[1] US Highways from US 1 to US 830 (http://www.us-highways.com/us1830.htm) Robert V. Droz

Browse numbered routes

External Links

- U.S. Route 278 in South Carolina (Mapmikey's South Carolina Highways Page) (http://www.angelfire.com/sc3/scroads/us278.html)

South_Carolina_Highway_417

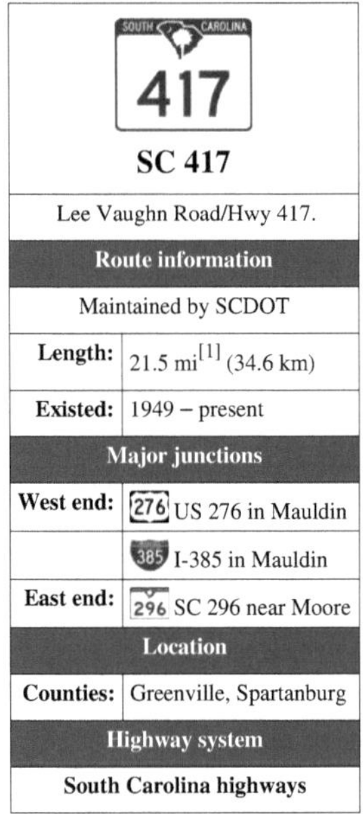

South Carolina Highway 417 is a numbered highway in Central Greenville and Spartanburg counties.

Route description

SC 417's western terminus is at junction with US 276 in Mauldin, SC and near the junctions of I-385 and I-185. Its overall course is in a northeasterly direction, and SC 417's eastern terminus is at a junction with SC 296 near Spartanburg, SC.

From SC 417's western terminus to its split from SC 14, SC 417 is known as Main St/Laurens Rd. From then until its merger with SC 146, SC 417 is known as Lee Vaughn Road. While it is merged with SC 146, SC 417 is known as Woodruff Rd. When it splits from SC 146, SC 417 is known as Highway 417, and it keeps that name until it reaches its eastern terminus.

Major intersections

County	Location	Mile	Destinations	Notes
Greenville	Mauldin		US 276	Western terminus
			I-385 (Freeway)	Exit 31 (I-385)
	Simpsonville		SC 14	Start SC 14 concurrency
			SC 14	End SC 14 concurrency
			SC 146	Start SC 146 concurrency
Spartanburg			SC 146	End SC 146 concurrency
	Cashville		SC 101	
			SC 290	
			SC 296	Eastern terminus

References

[1] Google, Inc. *Google Maps — overview map of SC Highway 417* (http://maps.google.com/maps?f=d&source=s_d&saddr=S+Main+St&
daddr=S+Carolina+417+S/E+Georgia+Rd+to:S+Carolina+417+S&
geocode=FaKeEgIdxRwY-w;FTpAEgIdNG4Z-w;FdBgFAIdVw8c-w&hl=en&mra=ls&sll=34.81378,-82.1765&sspn=0.27116,0.
441513&ie=UTF8&t=h&z=11) (Map). Cartography by Google, Inc. . Retrieved 2010-05-08.

Interstate_85_in_South_Carolina

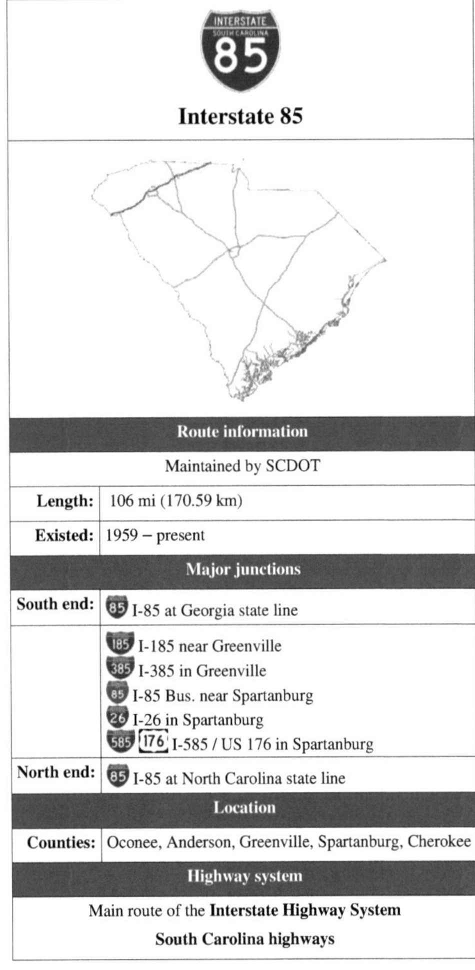

In the U.S. state of South Carolina, Interstate 85 runs northeast-southwest through Greenville and Spartanburg. It follows the general corridor of U.S. Route 29. In 1964, I-85 became the first South Carolina Interstate highway to have its originally planned mileage completed.

Route description

Georgia state line to Greenville

Interstate 85 North after Exit #1 in Oconee County, South Carolina in 2008.

In South Carolina, Interstate 85 crosses Lake Hartwell at several points before bypassing Anderson on the way to Greenville. Beginning at Anderson, I-85 widens from four to six lanes. Near Powdersville, U.S. 29 joins I-85 and they run concurrently until they cross the Saluda River.

Greenville to Spartanburg

Interstate 85 bypasses Greenville, but provides a link into the city via spur routes Interstates 185 and 385. U.S. 29 splits from I-85 and joins I-185 toward downtown Greenville. I-185 recently saw a major expansion south of the city as the Southern Connector was completed.

Two key Upstate businesses can be seen from this portion of the interstate. One is Michelin Tires's North American headquarters and the other is the BMW plant, located in Greer. I-85 also passes Greenville-Spartanburg International Airport, which serves the Greenville-Spartanburg metropolitan area.

Michelin North American headquarters in Greenville at Exit #54 in 2008.

Spartanburg to North Carolina state line

Instead of going into Spartanburg, I-85 bypasses the city to the north. Its original route is now designated Business Loop 85 and was approved by AASHTO on April 22, 1995.[1]

North of Spartanburg, I-85 narrows from six lanes back to four lanes and bypasses Gaffney. At Gaffney, motorists can see The Peachoid, a large water tower with its top shaped like a peach, representing one of the state's most important crops.

Most of the terrain between Spartanburg and the North Carolina border is rural in nature.

Related Routes

Auxiliary routes

- Interstate 185 is a spur into Greenville from the southwest, and a later extension as a toll road to I-385 southeast of Greenville.
- Interstate 385 is a spur into Greenville from the east, and a later extension southeast to Interstate 26 towards Columbia.
- Interstate 585 is a spur from Interstate 85 Business into Spartanburg from the northwest. It does not connect directly to I-85 or any other Interstate.

Business Route

- Business Loop 85 is a business Interstate route in Spartanburg.

Exit list

County	Location	#	Destinations	Notes
Oconee		1	SC 11 north (Cherokee Foothills Secenic Highway) – Walhalla	
		2	SC 59 north – Fair Play, Seneca, Oakway	
		4	SC 243 – Fair Play	
Anderson	Anderson	11	SC 24 to SC 243 – Anderson, Townville	
		14	SC 187 – Pendleton, Clemson	
		19	US 76 / SC 28 – Clemson, Anderson, Pendleton	Signed as exits 19A (east) and 19B (west)
		21	US 178 – Anderson, Liberty	
		27	SC 81 – Anderson	
		32	SC 8 – Pelzer, Easley, Belton	
		34	US 29 south – Williamston, Anderson	South end of US 29 overlap
		35	SC 86 – Piedmont, Easley	
	Easley	39	River Road – Easley	
		40	SC 153 – Easley	

County	Location	Exit	Destinations	Notes
Greenville		42	I-185 / US 29 north – Mauldin, Taylors	North end of US 29 overlap
	Greenville	44A	SC 20 (Piedmont Highway)	
		44B	US 25 (White Horse Road)	
		46A	Augusta Road	
		46B	SC 291 (Pleasantburg Drive)	
		46C	Mauldin Road	
		48	US 276 – Mauldin, Greenville	Signed as exits 48A (east) and 48B (west)
		51A	SC 146 (Woodruff Road)	
		51	I-385 – Mauldin, Greenville	Signed as exits 51B (south) and 51C (north)
		54	Pelham Road	
	Greer	56	SC 14 – Greer, Pelham	
Spartanburg		57	Aviation Drive – GSP International Airport	
		58	Brockman McClimon Road	
		60	SC 101 – Woodruff, Greer	
		63	SC 290 – Moore, Duncan	
		66	US 29 – Spartanburg, Wellford, Lyman	
		68	SC 129 west – Lyman	
		69	I-85 Bus. north – Spartanburg	
	Spartanburg	70	I-26 – Columbia, Asheville	Signed as exits 70A (east) and 70B (west)
		72	I-585 / US 176 – Spartanburg, Inman	Signed as exits 72A (south) and 72B (north)
		75	SC 9 – Spartanburg, Boiling Springs	
		77	I-85 Bus. south – Spartanburg	
		78	US 221 – Chesnee, Spartanburg	
		80	Road 57, Gossett Road	
		83	SC 110 – Cowpens, Chesnee	
Cherokee		87	Road 39	
	Gaffney	90	SC 105 south – Gaffney	
		92	SC 11 (Cherokee Foothills Scenic Highway) – Gaffney, Northgate	
		95	Road 82 – Gaffney	
		96	SC 18 – Gaffney, Shelby	
		98	Frontage Road	Northbound exit only
		100	Blacksburg Highway	
	Blacksburg	102	SC 5 / SC 198 – Blacksburg, Rock Hill	
		104	Road 99	
		106	US 29 – Blacksburg, Grover	

1.000 mi = 1.609 km; 1.000 km = 0.621 mi

Concurrency terminus · Closed/Former · Incomplete access · Unopened

References

[1] Interstate 85@Interstate-Guide.com (http://www.interstate-guide.com/i-085.html) Courtesy AARoads. Retrieved 27 May 2007.

Greenville_County,_South_Carolina

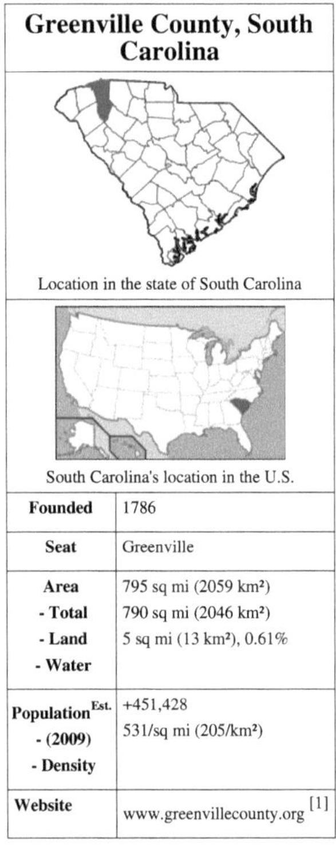

<table>
<tr><td colspan="2" align="center">Greenville County, South Carolina</td></tr>
<tr><td colspan="2" align="center">Location in the state of South Carolina</td></tr>
<tr><td colspan="2" align="center">South Carolina's location in the U.S.</td></tr>
<tr><td>Founded</td><td>1786</td></tr>
<tr><td>Seat</td><td>Greenville</td></tr>
<tr><td>Area
- Total
- Land
- Water</td><td>795 sq mi (2059 km²)
790 sq mi (2046 km²)
5 sq mi (13 km²), 0.61%</td></tr>
<tr><td>PopulationEst.
- (2009)
- Density</td><td>+451,428
531/sq mi (205/km²)</td></tr>
<tr><td>Website</td><td>www.greenvillecounty.org [1]</td></tr>
</table>

Greenville County is a county located in the state of South Carolina, United States. The population was 453,104 at the 2010 census, making it the most populous county in the state. It is included in the Greenville–Mauldin–Easley Metropolitan Statistical Area. Its county seat is the city of Greenville.[2]

Geography

According to the U.S. Census Bureau, the county has a total area of 795 square miles (2059.0 km^2), of which 790 square miles (2046.1 km^2) is land and 5 square miles (12.9 km^2) (0.61%) is water.[3]

Adjacent counties

- Henderson County, North Carolina - north
- Polk County, North Carolina - northeast
- Spartanburg County, South Carolina - east
- Laurens County, South Carolina - southeast
- Abbeville County, South Carolina - south
- Anderson County, South Carolina - southwest
- Pickens County, South Carolina - west
- Transylvania County, North Carolina - northwest

Demographics

As of the census[4] of 2010, there were 453,104 people, 198,546 households, and 201,997 families residing in the county. The population density was 497 people per square mile (186/km²). There were 186,803 housing units at an average density of 206 per square mile (80/km²). The racial makeup of the county was 77.53% White, 18.30% Black or African American, 0.19% Native American, 1.38% Asian, 0.05% Pacific Islander, 1.42% from other races, and 1.14% from two or more races. Hispanic or Latino of any race were 3.76% of the population. 19.6% were of American, 10.8% English, 9.2% German and 8.9% Irish ancestry according to Census 2000. 93.2% spoke English and 3.9% Spanish as their first language.

There were 149,556 households out of which 31.90% had children under the age of 18 living with them, 52.30% were married couples living together, 12.30% had a female householder with no husband present, and 31.80% were non-families. 26.80% of all households were made up of individuals and 8.50% had someone living alone who is 65 years of age or older. The average household size was 2.47 and the average family size was 3.00.

In the county the population was spread out with 25.60% of the population was under the age of 18, 9.60% from 18 to 24, 31.20% from 25 to 44, 25.80% from 45 to 64, and 13.70% who are 65 years of age or older. The median age was 36 years. For every 100 females, there were 94.80 males. For every 100 females age 18 and over, there were 91.60 males.

The median income for a household in the county was $42,049, and the median income for a family was $49,032. Males had a median income of $35,313 versus $27,034 for females. The per capita income for the county was $22,081. About 7.90% of families and 10.50% of the population were below the poverty line, including 13.20% of those under the age of 18 and 10.60% of those 65 and older.

Communities

The 2010 Census lists six cities and 16 census designated places that are fully or partially within Greenville County.[5] .

Cities

(population figures are from the 2000 census)

- Fountain Inn (pop. 6,317) (territory in Greenville and Laurens Counties)
- Greenville (pop. 61,802)
- Greer (pop. 16,843) (territory in Greenville and Spartanburg Counties)
- Mauldin (pop. 15,224)
- Simpsonville (pop. 14,352)
- Travelers Rest (pop. 4,099)

Unincorporated communities

- Berea
- City View
- Dunean
- Five Forks
- Gantt
- Golden Grove
- Judson
- Parker
- Piedmont (territory in Anderson and Greenville Counties)
- Sans Souci
- Slater-Marietta
- Taylors
- Tigerville
- Wade Hampton
- Ware Place
- Welcome

See also

- National Register of Historic Places listings in Greenville County, South Carolina

References

[1] http://www.greenvillecounty.org

[2] "Find a County" (http://www.naco.org/Counties/Pages/FindACounty.aspx). National Association of Counties. . Retrieved 2011-06-07.

[3] "US Gazetteer files: 2010, 2000, and 1990" (http://www.census.gov/geo/www/gazetteer/gazette.html). United States Census Bureau. 2011-02-12. . Retrieved 2011-04-23.

[4] "American FactFinder" (http://factfinder.census.gov). United States Census Bureau. . Retrieved 2008-01-31.

[5] See http://factfinder2.census.gov for population numbers and for municipality and CDP lists in the 2010 Census.

External Links

* Greenville Area Development Corporation (GADC) (http://www.greenvilleeconomicdevelopment.com/)

Hilton_Head_Island,_South_Carolina

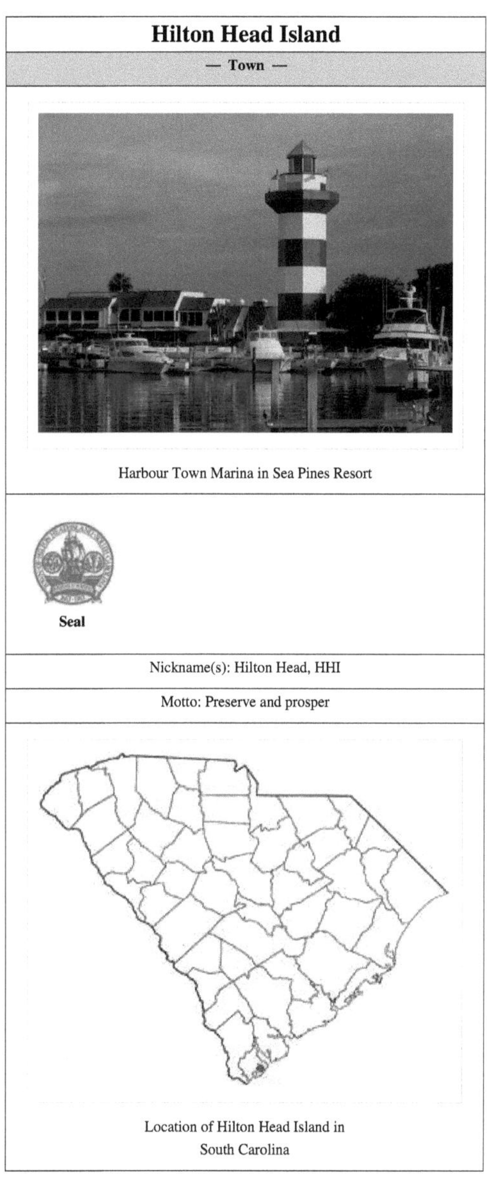

Hilton Head Island

— Town —

Harbour Town Marina in Sea Pines Resort

Seal

Nickname(s): Hilton Head, HHI

Motto: Preserve and prosper

Location of Hilton Head Island in
South Carolina

Coordinates: 32°10′44″N 80°44′35″W	
Country	United States
State	South Carolina
County	Beaufort
Incorporated (town)	1983
Government	
– **Mayor**	Drew Laughlin
– **Town manager**	Steve Riley
– **Fire chief**	Lavarn Lucas
Area	
– **Total**	55.5 sq mi (143.9 km^2)
– **Land**	42.1 sq mi (108.9 km^2)
– **Water**	13.5 sq mi (34.9 km^2) 24.28%
Elevation	10 ft (3 m)
Population (2010)	
– **Total**	37099
– **Density**	586.0/sq mi (310.8/km^2)
Time zone	EST (UTC-5)
– **Summer (DST)**	EDT (UTC-4)
ZIP code	29925, 29926, 29928
Area code(s)	843
FIPS code	45-34045[1]
GNIS feature ID	1246002[2]
Website	Official web site [3]

Hilton Head Island or **Hilton Head** is a resort town (located on an island of the same name) in Beaufort County, South Carolina, United States. It is 20 miles (32 km) north of Savannah, Georgia, and 95 miles (153 km) south of Charleston. The island gets its name from Captain William Hilton. In 1663, Captain Hilton identified a headland near the entrance to Port Royal Sound, which he named "Hilton's Head" after himself. The island features 12 miles (19 km) of beachfront on the Atlantic Ocean and is a popular vacation destination. In 2004, an estimated 2.25 million visitors pumped more than $1.5 billion into the local economy.[4] The year-round population was 37,099 at the 2010 census, although during the peak of summer vacation season the population can swell to 275,000.[5] Over the past decade, the island's population growth rate was 32%.[6]

The island has a rich history that started with seasonal occupation by native Americans thousands of years ago, and continued with European exploration and the Sea Island Cotton trade. It became an important base of operations for the Union blockade of the Southern ports during the Civil War. Once the island fell to Union troops, hundreds of ex-slaves flocked to Hilton Head, which is still home to many 'native islanders', many of whom are descendants of freed slaves known as the Gullah (or Geechee) who have managed to hold onto much of their ethnic and cultural identity.[7]

The Town of Hilton Head Island incorporated as a municipality in 1983 and is well known for its "eco-friendly" development.[8] The town's Natural Resources Division enforces the Land Management Ordinance which minimizes the impact of development and governs the style of buildings and how they are situated amongst existing trees.[9] As a result, Hilton Head Island enjoys an unusual amount of tree cover relative to the amount of development.[10] Approximately 70% of the island, including most of the tourist areas, is located inside gated communities.[11] However, the town maintains several public beach access points, including one for the exclusive use of town residents, who have approved several multi-million dollar land-buying bond referendums to control commercial growth.[12]

Hilton Head Island offers an unusual number of cultural opportunities for a community its size, including Broadway-quality plays at the Arts Center of Coastal Carolina, the 120-member full chorus of the Hilton Head Choral Society, the highly-rated Hilton Head Symphony Orchestra, the largest annual outdoor, tented wine tasting event on the east coast, and several other annual community festivals. It also hosts the Heritage Golf Classic, a stop on the PGA Tour which is played on the Harbour Town Golf Links in Sea Pines Resort.[13]

History

Early history

An ancient shell ring can be seen near the east entrance to the Sea Pines Forest Preserve. The ring, one of only 20 in existence, is 150 feet (46 m) in diameter and is believed to be over 15,000 years old. Archeologists believe that the ring was a refuse heap, created by Indians that lived in the interior of the ring, which was kept clear and used as a common area. Two other shell rings on Hilton Head were destroyed when the shells were removed and used to make tabby for roads and buildings. The Shell Ring is listed in the National Register of Historic Places and is protected by law.[14]

Baynard Mausoleum, oldest intact structure on HHI (built in 1846).

Since the beginning of recorded history in the New World, the waters around Hilton Head Island have been known, occupied and fought for in turn by the English, Spanish, French, and Scots.[15]

A Spanish expedition led by Francisco Cordillo explored the area in 1521, initiating European contact with local tribes.[16]

Fort Walker, Battle of Port Royal November 7, 1861.

In 1663, Captain William Hilton sailed on the *Adenture* from Barbados to explore lands granted by King Charles II of England to the eight Lords Proprietor. In his travels, he identified a headland near the entrance to Port Royal Sound. He named it "Hilton's Head" after himself.[17] He stayed for several days, making note of the trees, crops, "sweet water" and "clear sweet air".[18]

In 1698, Hilton Head Island was granted as part of a barony to John Bayley of Ballingclough, County of Tipperary, Kingdom of Ireland. Another John Bayley, son of the first, appointed Alexander Trench as the island's first retail agent. For a time, Hilton Head was known as Trench's Island. In 1729, Trench sold some land to John Gascoine which Gascoine named "John's Island" after himself. The land later came to be known as Jenkin's Island after another owner.[19]

In 1788, a small Episcopal church called the Zion Chapel of Ease was constructed for plantation owners. The old cemetery, located near the corner of William Hilton Parkway and Mathews Drive (Folly Field), is all that remains. Charles Davant, a prominent island planter during the Revolutionary War, is buried there.

He was shot by Captain Martinangel of Daufuskie Island in 1781.[17] It is also home to the oldest intact structure on Hilton Head Island, the Baynard Mausoleum, which was built in 1846.

William Elliott II of Myrtle Bank Plantation grew the first crop of Sea Island Cotton in South Carolina on Hilton Head Island in 1790.

Fort Walker was a Confederate fort in what is now Port Royal Plantation. The fort was a station for Confederate troops, and its guns helped protect the 2-mile (3 km) wide entrance to Port Royal Sound, which is fed by two slow-moving and navigable rivers, the Broad River and the Beaufort River. It was vital to the Sea Island Cotton trade and the southern economy.[20]

Dock built by Union troops on Hilton Head Island. April 1862.

On October 29, 1861, the largest fleet ever assembled in North America moved south to seize it.[21] In the Battle of Port Royal, the fort came under attack by the U.S. Navy, and on November 7, 1861, it fell to over 12,000 Union troops.[22] The fort would be renamed Fort Welles, in honor of Gideon Welles, the Secretary of the Navy.[23]

U.S. General Hospital, HHI. March 23, 1863.

Hilton Head Island would have tremendous significance in the Civil War, becoming an important base of operations for the Union blockade of the Southern ports, particularly Savannah and Charleston. The Union would also build a military hospital on Hilton Head Island with a 1200-foot (370 m) frontage and a floor area of 60000 square feet (6000 m^2).[24]

Hundreds of ex-slaves flocked to Hilton Head Island, where they could buy land, go to school, live in government housing, and serve in what was called the First Regiment of South Carolina Volunteers (although in the beginning, many were "recruited" at the point of a bayonet).[25] A community called Mitchelville (in honor of General Ormsby M. Mitchel) was constructed on the north end of the island to house them.[26]

The Leamington Lighthouse was built in the 1870s on the southern edge of what is now Palmetto Dunes.

On August 27, 1893, the Sea Islands Hurricane made landfall near Savannah, Georgia, with a storm surge of 16 feet (5 m) and swept north across South Carolina, killing over a thousand and leaving tens of thousands homeless.[27]

20th century

An experimental steam cannon guarding Port Royal Sound was built around 1900 in what is now Port Royal Plantation. The cannon was fixed but its propulsion system allowed for long range shots for the time.

In 1931, Wall Street tycoon, physicist, and patron of scientific research, Alfred Lee Loomis along with his brother-in-law and partner, Landon K. Thorne, purchased 17000 acres (69 km^2) on the island (over 63% of the total land mass) for about $120,000 to be used as a private game reserve.[28] [29]

Mitchelville "refugee quarters" 1864.

On the Atlantic coast of the island are large concrete gun platforms that were built to defend against a possible invasion by the Axis powers of World War II. Platforms like these can be found all along the eastern seaboard.

"The Beach Pounders" - U.S. Coast Guard Mounted Beach Patrol training on HHI during World War II.

The Mounted Beach Patrol and Dog Training Center on Hilton Head Island trained U.S. Coast Guard Beach Patrol personnel to use horses and dogs to protect the southeastern coastline of the U.S.[30]

In the early 1950s, three lumber mills contributed to the logging of 19000 acres (77 km^2) of the island.[16] The island population was only 300 residents.[16] Prior to 1956, access to Hilton Head was limited to private boats and a state-operated ferry. The island's economy centered on shipbuilding, cotton, lumbering, and fishing.[8]

The James F. Byrnes Bridge was built in 1956. It was a two-lane toll swing bridge constructed at a cost of $1.5 million that opened the island to automobile traffic from the mainland.[16] The swing bridge was hit by a barge in 1974 which shut down all vehicle traffic to the island until the Army Corps of Engineers built and manned a pontoon bridge while the bridge was being repaired. The swing bridge was replaced by the current four-lane bridge in 1982.[16]

The beginning of Hilton Head as a resort started in 1956 with Charles E. Fraser developing Sea Pines Resort. Soon, other developments followed, such as Hilton Head Plantation, Palmetto Dunes Plantation, Shipyard Plantation, and Port Royal Plantation, imitating Sea Pines' architecture and landscape. Sea Pines however continued to stand out by creating a unique locality within the plantation called Harbour Town, anchored by a recognizable lighthouse. Fraser was a committed environmentalist who changed the whole configuration of the marina at Harbour Town to save an ancient live oak.[11] It came to be known as the Liberty Oak, known to generations of children who

The Liberty Oak in Harbour Town

watched singer and song writer Gregg Russell perform under the tree for over 25 years.[31] Fraser was buried next to the tree when he died in 2002.[32]

The Heritage Golf Classic was first played in Sea Pines Resort in 1969, and has been a regular stop on the PGA Tour ever since.[13]

Also in 1969, the Hilton Head Island Community Association successfully fought off the development of a BASF chemical complex on the shores of Victoria Bluff (now Colleton River Plantation). Soon after, the association and other concerned citizens "south of the Broad" fought the development of off-shore oil platforms by Brown & Root (a division of Halliburton) and ten-story tall liquefied natural gas shipping spheres by Chicago Bridge & Iron.[33] These events helped to polarize the community, and the Chamber of Commerce started drumming up support for the town to incorporate as a municipality. After the Four Seasons Resort (now Hilton Head Resort) was built along William Hilton Parkway, a referendum of incorporation was passed in May 1983. Hilton Head Island had become a town.[33]

The Land Management Ordinance was passed by the Town Council in 1987. Disney's Hilton Head Island Resort opened in 1996. The Cross Island Parkway opened in January 1997. An indoor smoking ban in bars, restaurants, and public places took effect on May 1, 2007.[34]

Government

The Town of Hilton Head Island incorporated as a municipality in 1983 and has jurisdiction over the entire island except Mariner's Cove, Blue Heron Point, and Windmill Harbor.[5] The Town of Hilton Head Island has a Council-Manager form of government. The Town Manager is the chief executive officer and head of the administrative branch and is responsible to the municipal council for the proper administration of all the affairs of the town. The Town Council exercises all powers not specifically delegated to the Town Manager. The Mayor has the same powers, duties, and responsibilities as a member of Town Council. In addition, the Mayor establishes the agenda for Town Council meetings, calls special meetings, executes contracts, deeds, resolutions, and proclamations not designated to the Town Manager, and represents the town at ceremonial functions.[35]

Town departments include Building & Fire Codes, Business License, Code Enforcement, Finance, Fire & Rescue, Human Resources, Legal, Municipal Court, Planning, and Public Projects & Facilities.[36]

The town had a budget of $74,753,260 for fiscal year 2006/2007.[5] It consists of three separate fiscal accounting funds: the General Fund, the Capital Projects Fund, and the Debt Service Fund.[5] The General Fund is the operating fund for the town and accounts for all financial resources of the town except the Capital Projects Fund and the Debt Service Fund.[5] The Capital Projects Fund is used to acquire land and facilities, and improve public facilities, including roads, bike paths, fire stations, vehicle replacement, drainage improvements, and park development.[5] The Debt Service Fund accounts for the accumulation of resources and the payment of debt.[5] On Tuesday, June 5, 2007, the Town Council approved a $93,154,110 budget for fiscal year 2007/2008 on the first reading with a vote of 6-0.[37] The most recent budget, for the 2010/2011 fiscal year is $74,299,720 [38]

Office holders as of January 2011:

- Drew Laughlin, Mayor[39]
- Kenneth (Ken) S. Heitzke, Ward 6, Mayor Pro Tem[39]
- Willie (Bill) Ferguson, Ward 1[39]
- William (Bill) D. Harkins, Ward 2[39]
- Vacant, Ward 3[39]
- Kimberly (Kim) W. Likins, Ward 4[39]
- George W. Williams Jr., Ward 5[39]
- Stephen (Steve) Riley, Town Manager[39]

Council mission statement:

To provide excellent customer service to all that come in contact with the Town.

To wisely manage and utilize the financial and physical resources of Town government.

To promote policies and programs which will assure the long term health and vitality of the community.

To encourage and instill job satisfaction for all Town staff.

To develop and enhance the professional growth of all Staff members.[40]

Geography

Topography

Hilton Head Island is a shoe-shaped island that lies just north of Savannah, Georgia, and 90 miles (140 km) south of Charleston, South Carolina. The exact coordinates are 32°10′44″N 80°44′35″W (32.178828, -80.742947)[41].

According to the United States Census Bureau, the town has a total area of 55.5 square miles (144 km^2). Of that, 42.1 square miles (109 km^2) is land, and 13.5 square miles (35 km^2) or 24.28% is water.

Satellite image of Hilton Head Island. Accessed from NASA's World Wind project January 31, 2007.

Barrier island

Hilton Head Island is often referred to as the second largest barrier island on the eastern seaboard after Long Island (which is not actually a barrier island but two glacial moraines).[42] Technically, Hilton Head Island is only half barrier island. The north end of the island is a sea island dating to the Pleistocene epoch, and the south end is a barrier island that appeared as recently as the Holocene epoch. Broad Creek, which is actually a land-locked tidal marsh, separates the two halves of the island.[43]

The terrain of a barrier island is determined by a dynamic beach system with offshore bars, pounding surf, and shifting beaches; as well as grassy dunes behind the beach, maritime forests with wetlands in the interiors, and salt or tidal marshes on the lee side, facing the mainland. A typical barrier island has a headland, a beach and surf zone, and a sand spit.[44]

Climate

Culture

Arts Center of Coastal Carolina

Formerly the Self Family Arts Center, the Arts Center of Coastal Carolina is a showcase for professional performing and visual arts, as well as cultural festivals and educational outreach.[46] The Arts Center also offers community education, including Visual and Performing Arts Camps, Theater Camp, and other workshops and classes.[47]

Coastal Discovery Museum

The Coastal Discovery Museum, located at 100 William Hilton Parkway, offers a variety of programs, activities, and indoor and outdoor exhibits year-round to over 125,000 visitors. They include the Sea Island Classroom, the History Time-line Exhibit, and the Museum Store, plus 11 different tours and cruises around the island.[48] The Coastal Discovery Museum also leases the 68-acre (m^2) Honey Horn Plantation from the Town of Hilton Head Island, which it uses to educate residents and visitors about the history, culture and natural environment of the Lowcountry and Hilton Head Island.[49] The Coastal Discovery Museum is open Monday - Saturday, 9:00 a.m. - 5:00 p.m. and

Sunday 10:00 a.m. - 2:00 p.m.[50]

Hilton Head Choral Society

The Hilton Head Choral Society, founded in 1975,[51] is a non-profit organization "open to community members who love to sing and enjoy good fellowship."[51] The choirs of the Hilton Head Choral Society are known for their diverse musical repertoire[52] including classical masterworks, pops concerts and lighter fare, patriotic and Americana, gospel and musical theatre, a 20-voice chamber choir and a youth choir.[53] The 120-member full chorus presents four major programs per season: *A Fall Pops Concert*, *The Christmas Concert*, *The Musical Masterworks Concert* and a pair of Memorial Day concerts celebrating the art of American choral singing and a patriotic tribute.[51]

The Hilton Head Choral Society, full chorus.

Hilton Head Symphony Orchestra

The Hilton Head Symphony Orchestra was started 25 years ago by a handful of musicians and classical music aficionados who dreamed of bringing "big city" culture to Hilton Head.[54] Since then, they have transformed from a small group of classical music lovers to a highly rated symphony orchestra. Their main performance hall is the First Presbyterian Church on William Hilton Parkway, next to Fire Station 3.[55] A branch formed from the Hilton Head Symphony Orchestra is the Hilton Head Youth Orchestra, helping young musicians across the county with their love for music.

Arts Council of Beaufort County

ACBC's mission is to promote and foster the arts of Beaufort County, South Carolina, including Hilton Head Island. ACBC's vision is to position and maintain Beaufort County as a vibrant arts community and destination through active marketing, service to current arts organizations and artists and advocacy for the arts. ACBC programs include Quarterly Community Arts Grants, the Ever Expanding Arts Calendar, Get Your Art Out emerging artist initiative, the print publication ArtNews, and Arts of the Roundtables, which are free quarterly seminars exploring the business of art.[56]

Main Street Youth Theatre

The Main Street Youth Theatre, located on the north end of the island on Main Street, is a non-profit community theatre dedicated to enriching the lives of the islands youth by providing a true theatrical experience. Each year, MSYT performs 4-5 Broadway quality shows that run about 2 months at a time. During the tourist season, MSYT is a major tourist attraction and is also a local hot spot year round. The organization also provides acting, dance, and vocal instruction after school and during the summer.

Annual events

- **Gullah Celebration** - Although threatened by the rapid increase in tourism,[57] [58] Gullah culture can be seen at the annual Hilton Head Island Gullah Celebration which is held at Shelter Cove Community Park in February.[59] In the summer, the acclaimed Hallelujah Singers present a Gullah concert series at Hilton Head's Arts Center of Coastal Carolina.[7]

- **WineFest** - The 25th Annual WineFest was held in Shelter Cove Community Park on March 6, 2010.[60] [61] It is the largest outdoor, tented wine tasting on the East Coast, featuring over 1,500 domestic and international wines.[62] [63]

- **St. Patrick's Day Parade** - The 24th Annual St. Patrick's Day Parade was held on March 11, 2007. Joe Fraser, brother of Charles Fraser and former senior vice president of Sea Pines Plantation Co. was the grand marshal.[64] Over 20,000 people attended the parade, prompting Beaufort County Sheriff P.J. Tanner and other officials to question whether the parade might have outgrown its route along Pope Avenue.[65]

- **WingFest** - The 12th Annual Hargray WingFest was held at Shelter Cove Community Park on March 21, 2007. The event is operated by the Island Recreation Association, and all proceeds benefit the Island Recreation Scholarship Fund.[66]

- **HarbourFest** - HarbourFest,[67] now in its 19th season,[68] ran every Tuesday night from June 5 - August 21 at Shelter Cove Marina, featuring arts & crafts, live entertainment, and fireworks at sunset. There is a special HarbourFest celebration on July 4.[69] In addition, "Summertime at Shelter Cove" features five nights of family entertainment by Shannon Tanner, who plays two shows per night, 6:30 p.m. and 8:00 p.m., Monday - Friday.[70]

- **Heritage Golf Classic** - This is one of the busiest times of the year in HHI. A lot of famous people can be seen at the heritage. The 41st Annual Heritage Classic golf tournament took place April 14–20, 2010, at Harbour Town Golf Links in Sea Pines Resort.

- **Rib Burnoff & Barbecue Fest** - The 10th Annual Rib Burnoff & Barbecue Fest was held May 19, 2007 at Honey Horn Plantation.[71]

- **Celebrity Golf Tournament** - The 27th Annual Celebrity Golf Tournament was held August 31 - September 2, 2007, at the Golf Club at Indigo Run, the Robert Trent Jones course in Palmetto Dunes,[72] and Harbour Town Golf Links. To date the tournament has contributed over $2,785,000 to 18 children's charities.[73]

- **FoodFest** - FoodFest 2007 took place September 13–16, 2007. FoodFest celebrates the talent of the local hospitality industry and provides attendees with several spectator events including: The Best Bartender Drink Making Contest, The Hospitable Waiter's Race, and The Tailgate Gourmet Challenge.[74] [75]

- **Chili Cookoff** - The 23rd Annual Chili Cookoff was held Saturday, October 6, 2007, at 12:00 p.m. at Honey Horn Plantation.[76]

- **Community Festival** - The 7th Annual Community Festival took place Friday, October 19–20, 2007, from 5:30 - 9:30 p.m. at Honey Horn Plantation. It featured a "haunted trail" in the "haunted forest" presented by the Hilton Head Rotary Club and the Interact Clubs from Hilton Head High School and Hilton Head Prep.[77]

- **Concours d'Elegance & Motoring Festival** - The 6th Annual Concours d'Elegance & Motoring Festival, featuring world-class vintage and antique autos, Italian sports cars and American muscle cars from all over the country[78] was held November 1–4, 2007, at Honey Horn Plantation.[79]

- **The Dove Street Festival of Lights** - Started in 1990, takes place each December. More than 50 homes on Dove street decorate with holiday lights and the Glee Club of the Miami University of Ohio serenades residents with holiday songs. Town volunteers collect donations of money, food and toys at the festival that are given to The Deep Well Project, a local charity.[80]

Wildlife

Baby Loggerhead Sea Turtle

The Hilton Head Island area is home to a vast array of wildlife, including alligators, deer, Loggerhead Sea Turtles, manatee, hundreds of species of birds,[81] and dolphins.

The Coastal Discovery Museum, in conjunction with the South Carolina Department of Natural Resources, patrols the beaches from May through October as part of the Sea Turtle Protection Project.[82] The purpose of the project is to inventory and monitor nesting locations, and if necessary, move them to more suitable locations. During the summer months, the museum sponsors the Turtle Talk & Walk, which is a special tour designed to educate the public about this endangered species.[82] To protect Loggerhead Sea Turtles, a town ordinance stipulates that artificial lighting must be shielded so that it cannot be seen from the beach, or it must be turned off by 10:00 p.m. from May 1 to October 31 each year.[83]

The waters around Hilton Head Island are one of the few places on Earth where dolphins routinely use a technique called "strand feeding" whereby schools of fish are herded up onto mud banks, and the dolphins lie on their side while they feed before sliding back down into the water.[84] [85]

The saltmarsh estuaries of Hilton Head Island are the feeding grounds, breeding grounds, and nurseries for many saltwater species of game fish, sport fish, and marine mammals. The dense plankton population gives the coastal water its "murky" brown-green coloration.

Bottlenose dolphin in the Calibogue Sound just outside of Harbour Town Marina

Snowy egret with chicks

Plankton support marine life including oysters, shrimp and other invertebrates, and bait-fish species including Menhaden and Mullet, which in turn support larger fish and mammal species that populate the local waterways. Popular sport fish in the Hilton Head Island area include the Red Drum (or Spot Tail Bass), Spotted Sea Trout, Sheepshead, Cobia, Tarpon, and various shark species.[86]

Demographics

Historical populations			
Census	Pop.		%±
1990	23694		—
2000	33862		42.9%
2010	37099		9.6%
U.S. Decennial Census [87]			

As of the census[1] of 2000, there were 33,862 people, 14,408 households, and 9,898 families residing in the town, on a land area of 42.06 square miles (109 km^2). The population density was 805.1 people per square mile (310.8/km²). There were 24,647 housing units at an average density of 586.0 per square mile (226.3/km²).

Although the town occupies most of the land area of the island, it is not coterminous with it; there is a small part near the main access road from the mainland, William Hilton Parkway, which is not incorporated into the town. Hilton Head (the island) therefore has a slightly higher population (48,407 in Census 2000, defined as the Hilton Head Island Urban Cluster) and a larger land area (42.65 sq mi or 110.45 km²)

Location of the Hilton Head Island-Beaufort Micropolitan Statistical Area in South Carolina

than the town. The Hilton Head Island-Beaufort Micropolitan Area, which includes Beaufort and Jasper counties, had a 2005 estimated year-round population of 159,247.[88]

The racial makeup of the town was 85.33% White, 8.26% African American, 0.14% Native American, 0.55% Asian, 0.02% Pacific Islander, 4.48% from other races, and 1.21% from two or more races. Hispanic or Latino of any race were 11.48% of the population.

There were 14,408 households out of which 20.9% had children under the age of 18 living with them, 59.6% were married couples living together, 6.2% had a female householder with no husband present, and 31.3% were non-families. 23.8% of all households were made up of individuals and 11.3% had someone living alone who was 65 years of age or older. The average household size was 2.32 and the average family size was 2.68.

In the town the population was spread out with 17.3% under the age of 18, 6.9% from 18 to 24, 24.5% from 25 to 44, 27.3% from 45 to 64, and 24.1% who were 65 years of age or older. The median age was 46 years. For every 100 females there were 100.2 males. For every 100 females age 18 and over, there were 99.0 males.

Looking out over the tidal marsh to the Folly.

The median income for a household in the town was $60,438, and the median income for a family was $71,211. Males had a median income of $37,262 versus $30,271 for females. The per capita income for the town was $36,621. About 4.7% of families and 7.3% of the population were below the poverty line, including 9.4% of those under age 18 and 2.7% of those age 65 or over.

Hilton Head Island is part of the Hilton Head Island-Beaufort Micropolitan Statistical Area which includes Beaufort and Jasper counties and has a total estimated

2005 population of 159,247 (U.S. Census Bureau). According to the more detailed data available in the 2000 census, the population included in this micropolitan area (which actually was designated after the census itself) was 64% urban and 36% rural. It includes the urban clusters of Beaufort (2000 pop.: 46,128), Hilton Head Island (47,821), Bluffton (17,546), and Ridgeland (3,616). The urban clusters of Hilton Head Island and Bluffton will probably be merged by the 2010 Census.

Emergency services

Hilton Head Island Fire & Rescue began operations July 1, 1993 as a consolidation of the former Sea Pines Forest Beach Fire Department, the Hilton Head Island Fire District, and the Hilton Head Island Rescue Squad.[89] It is a career department that provides fire suppression and emergency medical services (EMS) at the advanced life support level. Special operations capabilities include HAZMAT, urban search and rescue (USAR), confined space rescue, trench rescue, and rope rescue. The department is accredited by the Commission on Fire Accreditation International (CFAI).[90] [91] [92]

Hilton Head Island Fire
& Rescue Emblem

There are seven fire stations on Hilton Head Island, providing professional fire protection and emergency medical care.

- Station 1: 70 Cordillo Parkway - (*in Shipyard Plantation near the Pope Avenue entrance*)[93]
- Station 2: 65 Lighthouse Road - (*in Sea Pines Resort between Frazer Circle and Harbour Town*)[93]
- Station 3: 534 William Hilton Parkway - (*across from Port Royal Plantation next to First Presbyterian Church*)[93]
- Station 4: 400 Squire Pope Road - (*near the back gate of Hilton Head Plantation*)[93]

Hilton Head Island Fire
& Rescue Patch

- Station 5: 20 Whooping Crane Way - (*near the front gate of Hilton Head Plantation*)[93]
- Station 6: 16 Queen's Folly Road - (*in the front of Palmetto Dunes under the water tower*)[93]
- Station 7: 1001 Marshland Road - (*by the toll booths of the Cross Island Parkway*)[93]
- Fire & Rescue Headquarters: 40 Summit Drive - (*general aviation entrance to the airport off Dillon Road, next to the convenience center*)

Hilton Head Island Fire & Rescue also works with Bluffton Township Fire Department as a sponsoring agency for two of South Carolina's designated special teams: one of the state's Hazardous Materials/Weapons of Mass Destruction Response Teams and one of the four Regional Urban Search and Rescue Response Teams.[94] [95] [96] [97]

Police services are contracted through Beaufort County Sheriff's Office.[5] The island is equipped with an enhanced 9-1-1 system.[89]

Gated communities

- Hilton Head Plantation
- Indigo Run Plantation
- Long Cove Plantation
- Palmetto Dunes Resort
- Palmetto Hall Plantation
- Port Royal Plantation
- Sea Pines Resort
- Shipyard Plantation
- Spanish Wells Plantation
- Wexford Plantation
- Windmill Harbour
- The Spa on Port Royal Sound

Public beach access

- Alder Lane Beach Access - 22 metered spaces[98]
- Burkes Beach Access - 13 metered spaces[98]
- Coligny Beach Park - parking is free - some parking reserved for annual beach passes from 8:00 p.m. to 3:00 p.m.[98]
- Driessen Beach Park - 207 long term parking spaces - some parking reserved for annual beach passes from 8:00 a.m. to 3:00 p.m.[98]
- Fish Haul Park - parking is free[98]
- Folly Field Beach Park - 51 metered spaces[98]
- Islanders Beach Park - annual beach pass parking only[98]
- Mitchelville Beach Park - parking is free[98]

Island parks

- Alder Lane Beach Access[99]
- Barker Field[99]
- Burkes Beach Access[99]
- Broad Creek Boat Ramp[99]
- Chaplin Community Park[99]
- Coligny Beach Park[99]
- Compass Rose Park[99]
- Cordillo Tennis Courts[99]
- Crossings Park & Bristol Sports Arena[99]
- Driessen Beach Park[99]
- Fish Haul Creek Park[99]
- Folly Field Beach Park[99]
- Green Shell Park[99]
- Hilton Head Park (Old Schoolhouse Park* Island Recreation Center[99]
- Islanders Beach Park[99]
- Jarvis Creek Park[99]
- Marshland Road Boat Landing[99]
- Old House Creek Dock[99]
- Shelter Cove Community Park[99]
- Xeriscape Garden[99]

Schools

Public schools

- Hilton Head Island Early Childhood Center (Pre K - K)
- Hilton Head Island School for the Creative Arts (Grades 1-5)
- Hilton Head Island International Baccalaureate Elementary School (Grades 1-5)
- Hilton Head Island Middle School
- Hilton Head Island High School

Private schools

- Hilton Head Preparatory School
- Hilton Head Christian Academy
- St. Francis Catholic School
- Heritage Academy
- Sea Pines Montessori Academy [100]

Notable residents

	Notability	Reference
Arthur Blank	owner NFL Atlanta Falcons & Home Depot, has a house in Sea Pines Resort	[101] [102]
Patricia Cornwell	fiction author	[103]
Bobby Cremins	NCAA men's basketball coach, currently resides in Charleston, but maintains a home in Hilton Head	[104] [105] [106]
Jim Ferree	American professional golfer who played on the PGA Tour and the Senior PGA Tour	[107]
Trevor Hall	singer-songwriter, was raised in Hilton Head	[108] [109]
Darrell Hedric	former head basketball coach at Miami University, former NBA head scout	
John Jakes	author of historical fiction, resides in Hilton Head	[102] [110]
Michael Jordan	former NBA player, sold his house on Hilton Head when his father died	[111]
John V. Lindsay	former mayor of New York City, died in Hilton Head on December 19, 2000	[112]
John Mellencamp	singer-songwriter	[102]
Mark Messier	NHL hockey player, part-time resident of Hilton Head	[113]
Serge Savard	former Montreal Canadiens defenseman and general manager	
Duncan Sheik	singer-songwriter, was raised in Hilton Head	[114] [115]
Stan Smith	former Wimbledon, US Open and Davis Cup Champ and Tennis Pro	[102]
Col. Benjamin H. Vandervoort	WWII hero, died in his home on Hilton Head in 1990 at the age of 75	[116]
Kathryn R. Wall	American author of mystery novels	[117]
Jayson Williams	former NBA Basketball player, owns a home on Hilton Head	[118]

In Popular Culture

In Big Trouble in Little Langley, an episode of American Dad!, Francine's birth parents Nick and Cassandra Dawson live there.

In the book By Order of the President, by W. E. B. Griffin, the President of the United States maintains a home on Hilton Head Island. This is where Charlie Castillo meets the President for the first time.

References

[1] "American FactFinder" (http://factfinder.census.gov). United States Census Bureau. . Retrieved 2008-01-31.

[2] "US Board on Geographic Names" (http://geonames.usgs.gov). United States Geological Survey. 2007-10-25. . Retrieved 2008-01-31.

[3] http://www.hiltonheadislandsc.gov/

[4] "Employment Fast Facts," (http://www.hiltonheadisland.org/chamber-of-commerce/relocation/employment-and-volunteer.asp) *Hilton Head Island - Bluffton Chamber of Commerce* - Accessed January 31, 2007.

[5] "Consolidated Municipal Budget Fiscal Year July 1, 2006 - June 30, 2007," (http://www.hiltonheadislandsc.gov/Depts/finance/budget/ FY2007Budget.pdf) *Town of Hilton Head Island*, Accessed July 6, 2007.

[6] "Lowcountry Workforce on the Web," (http://www.lcwow.com/demographics.htm) *Hilton Head Island - Bluffton Chamber of Commerce* - Accessed January 31, 2007.

[7] "Snapshots," (http://delta-sky.com/ginc/portrait/hilton_head_island/pdf/02_hhi_snapshots.pdf) *Sky Magazine (Delta)*, December 2007. Accessed December 24, 2007.

[8] William W. Starr, "Graceful Growth," (http://delta-sky.com/ginc/portrait/hilton_head_island/pdf/01_hhi_starressay.pdf) *Sky Magazine (Delta)*, December 2007. Accessed December 20, 2007.

[9] "20 Who Made a Difference," (http://www.lowcountrynow.com/features/20difference/krebsbig.shtml) *Lowcountry Now (Savannah Morning News), 2003* - Accessed February 16, 2007.

[10] Gale B. "Hilton Head: the canopy view," (http://findarticles.com/p/articles/mi_m1016/is_n11-12_v96/ai_9162894) *American Forests*, November–December 1990. Accessed February 16, 2007.

[11] "Hilton Head, way ahead of its time," (http://www.usatoday.com/travel/destinations/2006-08-31-hilton-head_x.htm) *USA Today, September 1, 2006* - Accessed February 14, 2007.

[12] Whitney T. and Gordon J. "An Investigation of Sprawl Development and Its Effect On Transportation Planning: The Lower Savannah Region of Government," (http://ntl.bts.gov/lib/9000/9700/9793/Whitney_2001_Final_Report.pdf) *South Carolina State University - School of Engineering Technology and Sciences, 2001* - Accessed February 15, 2007.

[13] "Hilton Head Island an unquestionable golfing mecca," (http://sportsline.com/golf/story/10117287) *CBS Sportsline.com*, April 9, 2007. Accessed May 8, 2007

[14] "The Indian Shell Ring" (http://web.archive.org/web/20080328193204/http://www.csaadmin.com/shell_ring.htm). Community Services Associates. Archived from the original on 28 March 2008. . Retrieved 11 November 2011.

[15] Carse R. (1981) *Hilton Head in the Civil War: Department of the South (20th Anniv. Ed.)*, p. 1. Columbia, SC: The State Printing Company, ISBN B000J6GUMC

[16] "A History Timeline of Hilton Head Island," (http://www.hiltonheadislandsc.gov/Island/history.html) *Town of Hilton Head Island Official Municipal Website*, Accessed July 6, 2007.

[17] "Reference Desk," (http://www.bcgov.net/lib_hh/referenc.htm) *Beaufort County Public Library - Hilton Head Island*, Accessed May 19, 2007.

[18] "Island History," (http://www.hiltonheadisland.com/history.htm) *HiltonHeadIsland.com*, Accessed May 19, 2007.

[19] Margaret Greer (1989) *The Sands of Time - A History of Hilton Head Island*, pp. 20-21. Hilton Head Island, SC: SouthArt, Inc., ISBN 0-9610698-2-1.

[20] Carse R. (1981) *Hilton Head in the Civil War: Department of the South (20th Anniv. Ed.)*, pp. 1-2. Columbia, SC: The State Printing Company, ISBN B000J6GUMC

[21] Carse R. (1981) *Hilton Head in the Civil War: Department of the South (20th Anniv. Ed.)* , p. 2. Columbia, SC: The State Printing Company, ISBN B000J6GUMC

[22] Carse R. (1981) *Hilton Head in the Civil War: Department of the South (20th Anniv. Ed.)*, p. 22. Columbia, SC: The State Printing Company, ISBN B000J6GUMC

[23] Carse R. (1981) *Hilton Head in the Civil War: Department of the South (20th Anniv. Ed.)*, p. 72. Columbia, SC: The State Printing Company, ISBN B000J6GUMC

[24] Carse R. (1981) *Hilton Head in the Civil War: Department of the South (20th Anniv. Ed.)*, p. 71. Columbia, SC: The State Printing Company, ISBN B000J6GUMC

[25] Carse R. (1981) *Hilton Head in the Civil War: Department of the South (20th Anniv. Ed.)*, p. 82. Columbia, SC: The State Printing Company, ISBN B000J6GUMC

[26] Carse R. (1981) *Hilton Head in the Civil War: Department of the South (20th Anniv. Ed.)*, p. 91. Columbia, SC: The State Printing Company, ISBN B000J6GUMC

[27] "The Sea Islands Hurricane" (http://www.victoriana.com/Hurricane/seaislandshurricane-1.html), *The Victorian Era Online*, Accessed May 19, 2007.

[28] Ragland R., "Landowner changed course of history," (http://www.lowcountrynow.com/stories/050302/LOCtuxedo.shtml/) *Carolina Morning News*, May 3, 2002. Accessed June 13, 2007.

[29] Herbert W., "Palace of Science," (http://www.prism-magazine.org/mar03/science.cfm) *Prism Magazine*, March 2003; Vol 20, No 7. Accessed June 13, 2007.

[30] Coast Guard Beach Patrol During World War II (http://www.uscg.mil/history/Beach_Patrol_Photo_Index.html), A Historic Photo Gallery, Accessed May 7, 2007.

[31] "Our Favorite Family Getaways" (http://www.parents.com/parents/story.jhtml?storyid=/templatedata/parents/story/data/6130.xml&categoryid=/templatedata/parents/category/data/1131554703863.xml&page=3), *Parents Magazine*, June 2005. Accessed May 8, 2007

[32] "Charles Fraser eulogized as great dreamer" (http://www.lowcountrynow.com/stories/122202/LOCfuneral.shtml), *Carolina Morning News*, December 22, 2002. Accessed May 8, 2007.

[33] "A Town is Born," (http://www.celebratehiltonhead.com/article/723/a-town-is-born) *Celebrate Hilton Head*, February 2008. Accessed February 2, 2008.

[34] Donnelly, Tim "Hilton Head approves smoking ban," (http://www.islandpacket.com/news/local/story/6378297p-5689297c.html) *The Island Packet, February 21, 2007* - Accessed February 21, 2007.

[35] "Town of Hilton Head Island Town Council and Manager" (http://www.hiltonheadislandsc.gov/Council/tcmain.html), *Town of Hilton Head Island Municipal Government Website*, Accessed July 6, 2007.

[36] "Departments" (http://www.hiltonheadislandsc.gov/depts.html), *Official Town of Hilton Head Island Municipal Government Website*, Accessed June 6, 2007.

[37] Donnelly, Tim "Hilton Head Island approves $93 million budget," (http://islandpacket.com/news/local/story/6541707p-5821554c.html) *The Island Packet*, June 6, 2007. Accessed June 6, 2007.

[38] (http://www.hiltonheadislandsc.gov/departments/finance/budget/FY2011Budget.pdf), *Town of Hilton Head Island FY 2011 Budget* Accessed January 24, 2011.

[39] "Town Council Members" (http://www.hiltonheadislandsc.gov/council/tcmembers.cfm), *Town of Hilton Head Island Municipal Government Website*, Accessed May 11, 2007.

[40] "Town of Hilton Head Island Mission Statement" (http://www.hiltonheadislandsc.gov/mission.html), *Official Town of Hilton Head Island Municipal Government Website*, Accessed May 13, 2007.

[41] "US Gazetteer files: 2010, 2000, and 1990" (http://www.census.gov/geo/www/gazetteer/gazette.html). United States Census Bureau. 2011-02-12. . Retrieved 2011-04-23.

[42] David Lauderdale, "We're the biggest barrier island, so why not try to be the best?" (http://www.islandpacket.com/news/local/story/6627105p-5903631c.html) *The Island Packet*, August 19, 2007. Accessed August 19, 2007.

[43] Ballantine T. (1991) *Tideland Treasures.* p. 19. Columbia, SC: University of South Carolina Press, ISBN 978-0872497955.

[44] Ballantine T. (1991) *Tideland Treasures.* p. 11. Columbia, SC: University of South Carolina Press, ISBN 0-87249-795.

[45] "Weatherbase: Historical Weather for Hilton Head Island, South Carolina, United States of America" (http://www.weatherbase.com/weather/weather.php3?s=961483&refer=). . Retrieved January 27, 2007.

[46] "Arts Center of Coastal Carolina," (http://artshhi.com/index.htm) *Arts Center of Coastal Carolina*, Accessed June 24, 2007.

[47] "Community Education," (http://artshhi.com/html/community_education_.html) *Arts Center of Coastal Carolina*, Accessed June 24, 2007.

[48] "Coastal Discovery Museum," (http://coastaldiscovery.org/) *Coastal Discovery Museum*, Accessed June 24, 2007.

[49] "Coastal Discovery Museum at Honey Horn events," (http://coastaldiscovery.org/pages/Discovery_Center.htm) *Coastal Discovery Museum*, Accessed June 24, 2007.

[50] "Museums," (http://www.hiltonheadisland.org/vacation-visitors-guide/what-to-see-and-do/arts-and-culture/museums.asp) *Hilton Head Island - Bluffton Chamber of Commerce*, Accessed June 26, 2007.

[51] "About Us," (http://www.hiltonheadchoralsociety.org/hiltonheadchoralsociety.asp) *The Hilton Head Choral Society*, Accessed June 27, 2007.

[52] "Love to sing? Join the choral society," (http://www.islandpacket.com/features/story/6541419p-5821319c.html) *The Island Packet*, June 6, 2007. Accessed June 27, 2007.

[53] "Welcome to the Hilton Head Choral Society - Celebrating 32 years of Beautiful Music," (http://www.hiltonheadchoralsociety.org/) *Hilton Head Choral Society*, Accessed June 27, 2007.

[54] "Hilton Head Symphony Orchestra Celebrates 25 Years," (http://www.hhso.org/about/history.html) *Hilton Head Symphony Orchestra*, Accessed June 24, 2007.

[55] "About the Symphony," (http://www.hhso.org/about/) *Hilton Head Symphony Orchestra*, Accessed June 24, 2007.

[56] "Arts Council Of Beaufort County," (http://www.beaufortcountyarts.com/acbcorg.html) *The Arts Council of Beaufort County*, Accessed August 15, 2007.

[57] "NPR Examines the Cost of Paradise Sea Island's Development and the Gullah Community" (http://www.npr.org/about/press/000828.gullah.html) *NPR, August 28, 2000* - Accessed February 16, 2007.

[58] "Effort to preserve Gullah and Geechee culture moves forward," (http://www.thestate.com/mld/thestate/news/local/16708990.htm) *The State, February 15, 2007.* Accessed February 16, 2007.

[59] "Gullah History," (http://www.gullahcelebration.com/) *Hilton Head Island Gullah Celebration*, Accessed May 26, 2007.

[60] "Hilton Head Area Hospitality Association 2007-2008 Festival Season," (http://www.hiltonheadhospitality.org/sections/festivals/) *Hilton Head Hospitality Association*, Accessed June 26, 2007.

[61] "Wine Fest 2008," (http://www.hiltonheadhospitality.org/sections/wine-fest/) *Hilton Head Hospitality Association*, Accessed June 25, 2007.

[62] "Hilton Head Island Winefestival, Tasting & Auction," (http://www.discoversouthcarolina.com/products/1975.aspx) *Discover South Carolina*, Accessed June 26, 2007.

[63] "Better with Age," (http://www.hiltonheadmonthlyarchives.com/archives/march_06/043-045.pdf) *Hilton Head Monthly*, March 2006. Accessed June 26, 2007.

[64] Donnelly, Tim "Fraser chosen as grand marshal of island's St. Patrick's Day parade," (http://islandpacket.com/news/local/story/6395193p-5703616c.html) *The Island Packet*, March 1, 2007. Accessed June 26, 2007.

[65] Donnelly, Tim "Island's parade outgrown its route?" (http://islandpacket.com/news/local/story/6423937p-5725857c.html) *The Island Packet*, March 20, 2007. Accessed June 26, 2007.

[66] Craig Hysell, "Wingin' It!" (http://www.celebratehiltonhead.com/article/426/wingin-it) *Celebrate Hilton Head*, Accessed June 25, 2007.

[67] HarbourFest at Palmetto Dunes Oceanfront Resort (http://www.palmettodunes.com/hilton-head-harbour-fest.php)

[68] "Summer's a Blast," (http://www.hiltonheadmonthlyarchives.com/archives/june_06/094-095.pdf) *Hilton Head Monthly*, June 2006. Accessed June 26, 2007.

[69] "Entertainment," (http://palmettodunes.com/hilton-head-dining.php) *Palmetto Dunes Resort*, Accessed June 26, 2007.

[70] "HarbourFest includes food, fireworks, fun," (http://www.islandpacket.com/features/story/6590887p-5868903c.html) *The Island Packet*, July 17, 2007. Accessed July 26, 2007.

[71] "Rib Burnoff at Honey Horn," (http://coastaldiscovery.org/pages/rib_burnoff.htm) *Coastal Discovery Museum*, Accessed June 25, 2007.

[72] Robert Trent Jones Course at Palmetto Dunes Oceanfront Resort (http://www.palmettodunes.com/activities/28th-Annual-HHI-Celebrity-Golf-Tournament)

[73] "Beneficiaries," (http://www.hhcelebritygolf.com/) *Hilton Head Island Celebrity Golf Tournament*, Accessed June 26, 2007.

[74] "FoodFest 2007," (http://www.hiltonheadhospitality.org/sections/food-fest/) *Hilton Hhead Hospitality Association*, Accessed June 25, 2007.

[75] "Food Fest," (http://www.hiltonheadisland.org/iebms/coe/coe_p2_details.aspx?eventid=6022&sessionid=fb1fb0fgnfa3fe0fdn) *Hilton Head Island - Bluffton Chamber of Commerce*, Accessed June 26, 2007.

[76] "2007's 23rd Annual Chili Cookoff," (http://www.hiltonheadkiwanis.org/Chili.htm) *Kiwanis Club of Hilton Head*, Accessed June 25, 2007.

[77] "Community Festival at Honey Horn," (http://coastaldiscovery.org/pages/community_Festival.htm) *Coastal Discovery Museum*, Accessed June 26, 2007.

[78] "Concours d'Elegance," (http://www.discoversouthcarolina.com/products/26487.aspx) *Discover South Carolina*, Accessed June 26, 2007.

[79] "Schedule of Events," (http://www.hhiconcours.com/eventinfo/scheduleofevents.asp) *Concours d'Elegance & Motoring Festival*, Accessed June 26, 2007

[80] islandpacket.com (http://www.islandpacket.com/opinion/columns/seafoam/story/1066762.html)

[81] "Birding Areas," (http://www.hiltonheadaudubon.org/birding_areas.htm) *Hilton Head Audubon Society*, Accessed June 28, 2007.

[82] "Sea Turtles on Hilton Head Island" (http://coastaldiscovery.org/pages/sea.htm), *Coastal Discovery Museum*, Accessed May 6, 2007.

[83] Town Ordinance on Sea Turtle Protection (http://coastaldiscovery.org/documents/sea_turtle_protection.pdf) Accessed May 6, 2007. Archived (http://web.archive.org/20070204102821/http://coastaldiscovery.org/documents/sea_turtle_protection.pdf) February 4, 2007 at the Wayback Machine

[84] "Coastal Stock(s) of Atlantic Bottlenose Dolphin: Status Review and Management," (http://www.nmfs.noaa.gov/pr/pdfs/species/coastalbottlenosestock.pdf) Proceedings and Recommendations from a Workshop held in Beaufort, North Carolina, 13–14 September 1993. U.S. Department of Commerce, National Oceanic and Atmospheric Administration, National Marine Fisheries Service. pp. 56-57.

[85] "National Geographic Television Exposes the Dark Side of Dolphins in New National Geographic Special 'Dolphins: The Wild Side'," (http://www.nationalgeographic.com/tv/press/990202.html) *NationalGeographic.com*, National Geographic Press Release, Show aired November 13, 1999, Retrieved November 13, 2007.

[86] "South Carolina Fishing," (http://www.atlanticanglers.com/states/south_carolina/index.html) *Atlantic Anglers*, Accessed June 28, 2007.

[87] http://www.census.gov/prod/www/abs/decennial/

[88] Hilton Head Island CCD vs. Hilton Head Island town, Beaufort County (http://factfinder.census.gov/servlet/DTTable?_bm=y&-show_geoid=Y&-tree_id=4001&-_caller=geoselect&-context=dt&-errMsg=&-all_geo_types=N&-mt_name=DEC_2000_SF1_U_P001&-redoLog=true&-transpose=N&-search_map_config=lb=50ll=enlt=4001lzf=0.0lms=sel_00decldw=0.018985283719497238ldh=0.012134324225769228ldt=gov.census.aff.domain.map.EnglishMapExtentlif=giflcx=-80.78090095722615lcy=32.21893955312739lzl=2lpz=2lbo=318:317:316:314:313:323:319lbl=362:393:358:357:356:355:354lft=350:349:335:389:388:332:331lfl=381:403:204:380:36 -PANEL_ID=p_dt_geo_map&-_lang=en&-geo_id=06000US4501391540&-geo_id=16000US4534045&-CONTEXT=dt&-format=&-search_results=16000US4534045&-ds_name=DEC_2000_SF1_U)

[89] "Hilton Head Island Fire & Rescue Division," (http://www.hiltonheadislandsc.gov/Depts/fire/firemain.html) *Town of Hilton Head Island Municipal Government Website* - Accessed January 31, 2007.

[90] 2002 IAFC Awards and Recognitions (http://info.jems.com/firerescue/exclus03/e0208k.html), *Fire Rescue Magazine*, Accessed May 8, 2007

[91] "Hilton Head Island Fire and Rescue Gets Top Honors," (http://www.wtoctv.com/Global/story.asp?S=7008517&nav=0qq6) *WTOC-TV News 11*, August 30, 2007. Accessed September 7, 2007.

[92] "Fire department earns accreditation," (http://www.islandpacket.com/news/briefs/story/6647426p-5923123c.html) *The Island Packet*, August 31, 2007. Accessed September 7, 2007.

[93] "Holiday Tour of Lights at the Hilton Head Island Fire Stations," (http://www.islandpacket.com/static/man/pdfs/20061120_Xmastour.pdf) *The Island Packet*, Accessed May 19, 2007.

[94] "Regional US&R Teams," (http://www.ffmob.sc.gov/index.asp?file=RegionalUSandR.htm) *South Carolina Firefighter Mobilization Oversight Committee*, Accessed May 20, 2007.

[95] "Safety First - The Hilton Head/Bluffton Disaster Response Team Trains for Emergency Rescue Missions," (http://bluffton-today.wehaa-server2.com/catalog.php?catalog_id=438) *The Bluffton Today*, January 10, 2008. Accessed January 12, 2008.

[96] "Firefighters in SC train in tornado scenario," (http://www.fireengineering.com/news/newsArticleDisplay.html?id=156017) *Fire Engineering*, January 9, 2008. Accessed January 12, 2008.

[97] "Mock tornado stirs up read training locally," (http://www.islandpacket.com/news/local/story/125943.html) *The Island Packet*, January 9, 2008. Accessed January 12, 2008.

[98] "Hilton Head Island Beaches," (http://www.hiltonheadislandsc.gov/Island/beaches.html) *Official Town of Hilton Head Island Municipal Government Website*, Accessed June 24, 2007.

[99] "Hilton Head Island Parks," (http://www.hiltonheadislandsc.gov/Maps/parksmap.pdf) *Official Town of Hilton Head Island Municipal Government Website*, Accessed June 24, 2007.

[100] http://www.spma.com

[101] "Sea Pines Resort - A Southern Slice of Heaven" (http://www.publinksgolfer.net/articles/365/1/bSea-Pines-Resort/Page1.html). *Golfing Magazine*. 2007-06-29. . Retrieved 2007-06-29.

[102] Berryman, Anne (2005-10-28). "An Enclave for Golfers, Beachgoers and Arts Lovers" (http://travel.nytimes.com/2005/10/28/realestate/28haven.html?fta=y). *The New York Times*. . Retrieved 2008-02-06.

[103] "They're baack: Every year, you can count on Ohio golfers heading to Hilton Head in droves" (http://www.golfohio.com/departments/features/hilton-head-travel.htm). *GolfOhio.com*. 2002-11-29. . Retrieved 2008-02-06.

[104] Garcia, Marlen (2006-11-06). "Cremins is unretiring with passion" (http://www.usatoday.com/sports/college/mensbasketball/2006-11-06-cremins-cover_x.htm). *USA Today*. . Retrieved 2008-02-06.

[105] "Cremins Jettisons Life of Leisure" (http://www.washingtonpost.com/wp-dyn/content/article/2008/01/26/AR2008012601978.html). *The Washington Post*. 2008-01-27. . Retrieved 2008-02-06.

[106] Bernstein, Viv (2006-11-28). "Back on the Court, Cremins Has a New Bounce" (http://www.nytimes.com/2006/11/28/sports/ncaabasketball/28cremins.html). *The New York Times*. . Retrieved 2006-11-28.

[107] "Ferree is at Home at Long Cove Club" (http://www.uswmidam.org/2003/results/Ferree_story.html). .

[108] "Interview - Trevor Hall" (http://www.newbeats.com/2006/03/interview-trevor-hall.html). *NewBeats*. . Retrieved 2008-02-06.

[109] "Trevor Hall: A young rascal with a bright recording future" (http://findarticles.com/p/articles/mi_qn4176/is_20060127/ai_n16039785). *Oakland Tribune reprinted at CNET's findarticles.com*. 2006-01-27. . Retrieved 2008-02-06.

[110] "Acclaimed Hilton Head author gets S.C. award" (http://www.islandpacket.com/266/story/227406.html). *The Island Packet*. 2008-02-15. . Retrieved 2008-02-15.

[111] Applebome, Peter (1994-09-02). "Tourism Enriches an Island Resort, But Hilton Head Blacks Feel Left Out" (http://query.nytimes.com/gst/fullpage.html?res=9407E1DA1438F931A3575AC0A962958260). *The New York Times*. . Retrieved 2008-02-06.

[112] "Paid Notice: Deaths LINDSAY, JOHN V." (http://query.nytimes.com/gst/fullpage.html?res=9D05E5D7163EF937A15751C1A9669C8B63). *The New York Times*. 2000-12-24. . Retrieved 2008-02-06.

[113] "Island resident Mark Messier reflects on Hockey Hall of Fame induction" (http://web.archive.org/web/20071109112816/http://www.islandpacket.com/sports/local/story/73699.html). *The Island Packet*. 2007-11-08. Archived from the original (http://www.islandpacket.com/sports/local/story/73699.html) on November 9, 2007. . Retrieved 2007-11-08.

[114] Bellafante, Ginia (2006-12-26). "Broadway Is Rocker's Latest Alternative" (http://www.nytimes.com/2006/12/26/theater/26shei.html). *The New York Times*. . Retrieved 2008-02-06.

[115] "And the Hilton Head Grammy winner is ...." (http://www.islandpacket.com/266/story/227397.html). *The Island Packet*. 2008-02-15. . Retrieved 2008-02-06.

[116] "Local WWII hero -- portrayed by John Wayne in "The Longest Day" -- will be honored today" (http://web.archive.org/web/20070927235044/http://islandpacket.com/news/local/story/6532724p-5813010c.html). *The Island Packet*. 2007-05-28. Archived from the original (http://islandpacket.com/news/local/story/6532724p-5813010c.html) on September 27, 2007. . Retrieved 2007-05-28.

[117] "Welcome to Kathryn Wall's website" (http://www.kathrynwall.com/). . Retrieved 2008-05-24.

[118] "24. Bad Blood" (http://espn.go.com/magazine/vol5no16williams.html). *ESPN The Magazine*. 2007-07-25. . Retrieved 2008-02-06.

External links

- Town website (http://www.hiltonheadislandsc.gov)
- The Island Packet - local newspaper (http://www.islandpacket.com/)
- Hilton Head Island Chamber of Commerce (http://www.hiltonheadisland.org/chamber-of-commerce/)
- Hilton Head Gullah Festival (http://www.visitsouthcarolina.net/gullah-festival-2010)
- A Hilton Head Travel Guide (http://www.hhisleinfo.com/)
- Comprehensive Hilton Head Island visitor's site (http://www.hiltonheadview.com/)
- Complete Hilton Head Vacations Guide (http://www.hiltonheadvacationsguide.com/)

Simpsonville,_South_Carolina

<table>
<tr><td colspan="2" align="center">Simpsonville, South Carolina</td></tr>
<tr><td colspan="2" align="center">— City —</td></tr>
<tr><td colspan="2" align="center">Location of Simpsonville, South Carolina</td></tr>
<tr><td colspan="2" align="center">Coordinates: 34°44′0″N 82°15′36″W</td></tr>
<tr><td>Country</td><td>United States</td></tr>
<tr><td>State</td><td>South Carolina</td></tr>
<tr><td>County</td><td>Greenville</td></tr>
<tr><td>Government</td><td></td></tr>
<tr><td> – Mayor</td><td>Dennis C. Waldrop</td></tr>
<tr><td>Area</td><td></td></tr>
<tr><td> – Total</td><td>6.2 sq mi (16.1 km^2)</td></tr>
<tr><td> – Land</td><td>6.2 sq mi (16.1 km^2)</td></tr>
<tr><td> – Water</td><td>0.0 sq mi (0.0 km^2)</td></tr>
<tr><td>Elevation</td><td>860 ft (262 m)</td></tr>
<tr><td>Population (2000)</td><td></td></tr>
<tr><td> – Total</td><td>14352</td></tr>
<tr><td> – Density</td><td>2306.1/sq mi (890.4/km^2)</td></tr>
<tr><td>Time zone</td><td>Eastern (EST) (UTC-5)</td></tr>
<tr><td> – Summer (DST)</td><td>EDT (UTC-4)</td></tr>
<tr><td>ZIP codes</td><td>29680-29681</td></tr>
<tr><td>Area code(s)</td><td>864</td></tr>
<tr><td>FIPS code</td><td>45-66580[1]</td></tr>
<tr><td>GNIS feature ID</td><td>1250898[2]</td></tr>
<tr><td>Website</td><td>www.simpsonville.com [3]</td></tr>
</table>

Simpsonville is a city in Greenville County, South Carolina, United States. It is part of the Greenville–Mauldin–Easley Metropolitan Statistical Area. The population was 14,352 at the 2000 census, and estimated at 17,778 in 2009. Simpsonville is part of the "Golden Strip", along with Mauldin and Fountain Inn, which is noted for having low unemployment due to a diversity of industries including Para-Chem, Kemet, and Milliken. Simpsonville is home to Hillcrest High School. Simpsonville is also home to the 2008 Little League Softball World Champions.

Geography

Simpsonville is located at 34°44′0″N 82°15′36″W (34.733375, -82.260001)[4] . Elevation 860 ft. The city is located between Mauldin, South Carolina and Fountain Inn along Highway 14 and Interstate 385.

According to the United States Census Bureau, the city has a total area of 6.2 square miles (16.1 km²), all of it land.

Demographics

As of the census[1] of 2000, there were 14,352 people, 5,391 households, and 4,021 families residing in the city. The population density was 2,306.1 people per square mile (890.9/km²). There were 5,636 housing units at an average density of 905.6 per square mile (349.9/km²). The racial makeup of the city was 82.69% White, 13.76% African American, 0.23% Native American, 0.79% Asian, 0.04% Pacific Islander, 1.23% from other races, and 1.27% from two or more races. Hispanic or Latino of any race were 4.65% of the population.

There were 5,391 households out of which 39.9% had children under the age of 18 living with them, 58.6% were married couples living together, 11.9% had a female householder with no husband present, and 25.4% were non-families. 20.9% of all households were made up of individuals and 4.9% had someone living alone who was 65 years of age or older. The average household size was 2.66 and the average family size was 3.09.

In the city the population was spread out with 28.4% under the age of 18, 8.2% from 18 to 24, 34.4% from 25 to 44, 21.6% from 45 to 64, and 7.5% who were 65 years of age or older. The median age was 34 years. For every 100 females there were 97.0 males. For every 100 females age 18 and over, there were 94.6 males.

The median income for a household in the city was $47,223, and the median income for a family was $52,043. Males had a median income of $38,680 versus $26,394 for females. The per capita income for the city was $21,139. About 5.2% of families and 6.1% of the population were below the poverty line, including 6.8% of those under age 18 and 6.1% of those age 65 or over.

Crime

Simpsonville has a large appeal to middle class families. While its statistics are higher than national average crime rates, they are drastically lower than major U.S. cities. Violent crimes in 1995 totalled 22. Almost 10 years later, in 2004, they totalled 98 for the year. 2006 statistics of violent crime in Simpsonville reflect there was not one murder, reported incidents of rape were slightly higher than the national average(39 per 100,000 in Simpsonville, with a National Average of 33 per 100,000), and incidents of aggravated assaults were what tipped the 2006 violent crime scales, tallying in at 75% over the National Average. [5] In 2007 the personal crime incidents rate tallied in at 6 per 1000 residents, while the national average was 1.3 per 1000.[6] In September 2007 the FBI reported that the State of South Carolina's violent crime rate was the highest in the nation per capita, although Simpsonville is not mentioned at all in the article.[7] Car theft in Simpsonville was lower than the national average, causing unsubstantiated claims that Simpsonville allegedly has a high rate of car-jackings to be discredited. [5]

Government

Simpsonville is governed by a Mayor, a City Council, and several boards and commissions. The current officeholders are:

* Mayor: Dennis C. Waldrop (5th term ends 12-31-2011, to be replaced by Perry Eichor or Tammy Bagwell pending the results of a runoff election.)
* City Administrator: Russell Hawes
* Council Ward I: Ronald Bridges (Mayor Pro Tempore)(Current term ends 12-31-2011, to be replaced by Matthew Gooch)
* Council Ward II: Brown Garrett (2nd term ends 12-31-2013)
* Council Ward III: Patty Sanders (1st term ends 12-31-2011, to be replaced by Geneva Lawrence)
* Council Ward IV: Julius Welborn, M.D. (1st term ends 12-31-2013)
* Council Ward V: George Curtis (2nd term ends 12-31-2015)
* Council Ward VI: Sylvia Lockaby (1st term ends 12-31-2013)

Notable residents

* Jason Bokar, Chess Grandmaster
* Danelle German, creator of cat handbags and founder of the National Cat Groomers Institute of America.
* Shane Hall, NASCAR driver
* Tommy Jones, Professional bowler; 2005-06 PBA Player of the Year
* Travelle Wharton, offensive tackle for the Carolina Panthers.
* Jamon Meredith, Offensive Tackle and 2009 5th Round Draft Pick of the Green Bay Packers
* Lucas Glover, PGA Tour golfer and winner of the 2009 U.S. Open Golf Championship.

References

[1] "American FactFinder" (http://factfinder.census.gov). United States Census Bureau. . Retrieved 2008-01-31.

[2] "US Board on Geographic Names" (http://geonames.usgs.gov). United States Geological Survey. 2007-10-25. . Retrieved 2008-01-31.

[3] http://www.simpsonville.com/

[4] "US Gazetteer files: 2010, 2000, and 1990" (http://www.census.gov/geo/www/gazetteer/gazette.html). United States Census Bureau. 2011-02-12. . Retrieved 2011-04-23.

[5] http://simpsonville.areaconnect.com/crime1.htm

[6] MONEY Magazine: Best places to live 2007: Simpsonville, SC snapshot (http://money.cnn.com/magazines/moneymag/bplive/2007/snapshots/PL4566580.html)

[7] S.C.'s violent crime rate again highest in nation | Spartanburg, South Carolina | GoUpstate.com (http://www.goupstate.com/apps/pbcs.dll/article?AID=2007709260329&source=email), Spartanburg Herald-Journal

External links

* City of Simpsonville (http://www.simpsonville.com)

Mauldin,_South_Carolina

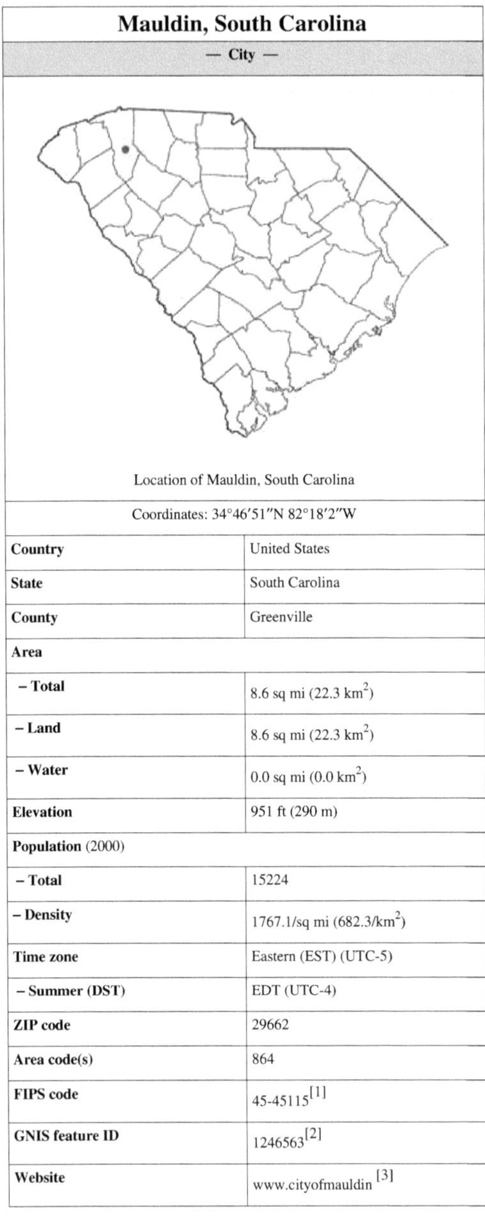

<table>
<tr><td colspan="2" align="center">Mauldin, South Carolina
— City —</td></tr>
<tr><td colspan="2" align="center">Location of Mauldin, South Carolina</td></tr>
<tr><td colspan="2" align="center">Coordinates: 34°46′51″N 82°18′2″W</td></tr>
<tr><td>Country</td><td>United States</td></tr>
<tr><td>State</td><td>South Carolina</td></tr>
<tr><td>County</td><td>Greenville</td></tr>
<tr><td>Area</td><td></td></tr>
<tr><td>– Total</td><td>8.6 sq mi (22.3 km^2)</td></tr>
<tr><td>– Land</td><td>8.6 sq mi (22.3 km^2)</td></tr>
<tr><td>– Water</td><td>0.0 sq mi (0.0 km^2)</td></tr>
<tr><td>Elevation</td><td>951 ft (290 m)</td></tr>
<tr><td>Population (2000)</td><td></td></tr>
<tr><td>– Total</td><td>15224</td></tr>
<tr><td>– Density</td><td>1767.1/sq mi (682.3/km^2)</td></tr>
<tr><td>Time zone</td><td>Eastern (EST) (UTC-5)</td></tr>
<tr><td>– Summer (DST)</td><td>EDT (UTC-4)</td></tr>
<tr><td>ZIP code</td><td>29662</td></tr>
<tr><td>Area code(s)</td><td>864</td></tr>
<tr><td>FIPS code</td><td>45-45115[1]</td></tr>
<tr><td>GNIS feature ID</td><td>1246563[2]</td></tr>
<tr><td>Website</td><td>www.cityofmauldin [3]</td></tr>
</table>

Mauldin is a city in Greenville County, South Carolina, United States. The population was 15,224 at the 2000 census. It is a principal city of the Greenville–Mauldin–Easley Metropolitan Statistical Area.

Geography

According to the United States Census Bureau, the city has a total area of 8.6 square miles (22 km^2), all of it land.

History

Benjamin Griffith was awarded the first land grant in what is now called Mauldin in 1784. The name of Mauldin was given to the town almost accidentally in 1820 thanks to South Carolina's Lieutenant Governor, W. L. Mauldin. The train station was called "Mauldin" because the Lt. Governor had assisted in getting the Greenville Laurens Railroad Company to come through the village. Over time, the entire area took the name of Mauldin.

During the Civil War, many of Mauldin's citizens left to fight and the city virtually dried up. It never completely recovered until after World War II when the city was once again incorporated as a town (1960).

Demographics

As of the census[1] of 2000, there were 15,224 people, 6,131 households, and 4,242 families residing in the city. The population density was 1,767.1 people per square mile (681.9/km²). There were 6,500 housing units at an average density of 754.5 per square mile (291.1/km²). The racial makeup of the city was 74.25% White, 20.82% African American, 0.30% Native American, 2.24% Asian, 0.11% Pacific Islander, 0.98% from other races, and 1.31% from two or more races. Hispanic or Latino of any race were 2.73% of the population.

There were 6,131 households out of which 33.9% had children under the age of 18 living with them, 55.8% were married couples living together, 10.4% had a female householder with no husband present and 30.8% were non-families. 26.0% of all households were made up of individuals and 6.2% had someone living alone who was 65 years of age or older. The average household size was 2.46 and the average family size was 2.97.

In the city the population was spread out with 25.0% under the age of 18, 8.1% from 18 to 24, 33.5% from 25 to 44, 24.1% from 45 to 64, and 9.3% who were 65 years of age or older. The median age was 35 years. For every 100 females there were 93.2 males. For every 100 females age 18 and over, there were 91.0 males.

The median income for a household in the city was $51,657, and the median income for a family was $61,817. Males had a median income of $41,047 versus $29,985 for females. The per capita income for the city was $24,750. About 3.2% of families and 4.4% of the population were below the poverty line, including 6.7% of those under age 18 and 9.2% of those age 65 or over.

School

The only high school is the Mauldin High School.

Notable natives

- Kevin Garnett - Olympian, professional basketball player
- Orlando Jones - actor

References

[1] "American FactFinder" (http://factfinder.census.gov). United States Census Bureau. . Retrieved 2008-01-31.
[2] "US Board on Geographic Names" (http://geonames.usgs.gov). United States Geological Survey. 2007-10-25. . Retrieved 2008-01-31.
[3] http://www.cityofmauldin.org/

External links

- City of Mauldin (http://www.cityofmauldin.org/)
- Mauldin High School (http://www.greenville.k12.sc.us/mauldinh/)

Article Sources and Contributors

Interstate_185_(South_Carolina) *Source*: http://en.wikipedia.org/w/index.php?title=Interstate_185_%28South_Carolina%29 *Contributors*: Amphytrite, Artisol2345, Bdj95, C.Fred, Carolina wren, Dough4872, Freewayguy, GETONERD84, Georgia guy, I-10, Imzadi1979, Kamlung, Lewisistheone1991, Mb29680, Modster, Morriswa, NE2, Nyttend, O, OrangeCounty39, Paddu, Polaron, R'n'B, Rschen7754, SPUI, Sam8, Shark96z, Tstarl0425, TwinsMetsFan, Vegaswikian, WhosAsking, 67 anonymous edits

South_Carolina_Department_of_Transportation *Source*: http://en.wikipedia.org/w/index.php?title=South_Carolina_Department_of_Transportation *Contributors*: Alansohn, CultureDrone, DH85868993, Elassint, Epbr123, JHunterJ, Kumioko, LilHelpa, Maddie!, Ohconfucius, Overpush, PennySpender1983, Rjwilmsi, Scgatorfan, Tassedethe, 10 anonymous edits

U.S._Route_29 *Source*: http://en.wikipedia.org/w/index.php?title=U.S._Route_29 *Contributors*: 8th Ohio Volunteers, Atanamir, Azitnay, Blahblah29, CC21560, Cleduc, ClovisPt, Crackerjackal, DavidLevinson, Denelson83, DidacticRogue, Dkastner, Doctor Whom, Dough4872, EaglesFanInTampa, Egdeltar, Elf, Engineer Bob, Fredddie, Gyrofrog, Henrymrx, Iain99, Imzadi1979, Jeff02, Jhb 10s, John Foxe, John K, Keeper76, Khatru2, Ljthefro, Lostcatholic, Lpangelrob, Ltljltlj, MPD01605, Mild Bill Hiccup, Morriswa, NE2, Niteowlneils, O, Obteene, Onore Baka Sama, Perimeter285, Polaron, Q300r bc2, Rafiguroian, Rfc1394, Rschen7754, Rufus843, SPUI, SchuminWeb, Scott5114, Stanton49, Stratosphere, Tariqabjotu, Tavy10, Terrek, TheOneKEA, Tom, Triadian, Tstarl0425, TwinsMetsFan, UNCCTF, Ubermonkey, VT hawkeye, Vegaswikian, WOSlinker, Washuotaku, WillC, Zoicon5, 103 anonymous edits

Interstate_385 *Source*: http://en.wikipedia.org/w/index.php?title=Interstate_385 *Contributors*: Bdj95, Cyndaquazy, DVac525, DanTD, DuncanHill, Freewayguy, GETONERD84, I-10, Imzadi1979, JamieS93, Jtm, JustAGal, Kamlung, Koavf, KyleAndMelissa22, Lewisistheone1991, Mazerunner, Morriswa, Nyttend, O, OrangeCounty39, Paddu, Pepper, Polaron, Radiojon, Rschen7754, Rufus843, SPUI, Sct72, Synchronism, Tedickey, Tstarl0425, TwinsMetsFan, Vchimpanzee, WhisperToMe, WillC, 112 anonymous edits

Interstate_85 *Source*: http://en.wikipedia.org/w/index.php?title=Interstate_85 *Contributors*: A86128, Airtuna08, Akhenaton06, All in, Andrewnpeters, Anthony, AnthonyA7, Auburnirishman, AudiR10, Bdj95, Bkonrad, Bluemoose, Bobblewik, Bodo920, Brennanrobison, Bryan Derksen, C.Fred, CC21560, CNCGuy, Cleduc, Creidieki, Cyclonenim, DVac525, Dabby, DanceBoatsi, DavidSteinle, Dcoetzee, Denelson83, Docu, Fratrep, GETONERD84, GT, Golbez, Gooday.1, Gpietsch, H7924, Hmains, Horologium, Jackhamma, Jackthomas, Jameslove, Jamesontai, Jedge242, Jevansen, Jimwilliams57, Jjc104, Jkeene, JohnnyAlbert10, Jpo, Jusjih, Kashaud17, Khatru2, Kimyou, Koavf, L1751207, LarryMac, Littledrummrboy, MPD01605, Marcika, Marvin01, Mat cross, Mathpianist93, Mcmillen76, MeltBanana, Mhking, Michael Patrick, Micman2b, Miller17CU94, Moabdave, Morriswa, Mosmof, MuzikJunky, MyReference, NE2, News2add, Nick Number, Noveltyghost, Nyttend, O, PGPirate, Paddu, Polaron, Qdub4lyfe, Radiojon, Radius, RandorXeus, Rfc1394, Rich Farmbrough, Roadawg, Roaddawg, Robertb-dc, Rschen7754, SPUI, Sbeath, Scott5114, Simplex1swrhs, Sonrisasgrandes, Spell4yr, Splat5572, Suodrak, Synchronism, Syunit, Tckma, Tim1357, Trey, Triadian, Truthanado, Tstarl0425, TwinsMetsFan, Vaoverland, Washuotaku, Whhalbert, WhisperToMe, WhosAsking, Will Beback, WillC, XinaNicole, Zoicon5, Zzyzx11, 179 anonymous edits

U.S._Route_278 *Source*: http://en.wikipedia.org/w/index.php?title=U.S._Route_278 *Contributors*: 25or6to4, Alexwcovington, Bdj95, Brandonrush, Bryant234, CC21560, Chris31416, Cleduc, Dale101usa, DanTD, Darklilac, Dough4872, Dr Debug, Fingers-of-Pyrex, Fredddie, Imzadi1979, Jkh8502, Kacie Jane, MPD01605, Mhking, Mjbyars1, NE2, Ospalh, Plastikspork, Radiojon, RIquall, RobyWayne, Rschen7754, SPUI, Southernroads, Synergy, TampAGS, TwinsMetsFan, Ubermonkey, WOSlinker, Wpgeek431, Xnatedawgx, Zoicon5, 43 anonymous edits

South_Carolina_Highway_417 *Source*: http://en.wikipedia.org/w/index.php?title=South_Carolina_Highway_417 *Contributors*: Cyndaquazy, Imzadi1979, R'n'B, Washuotaku, Xezbeth, 16 anonymous edits

Interstate_85_in_South_Carolina *Source*: http://en.wikipedia.org/w/index.php?title=Interstate_85_in_South_Carolina *Contributors*: Armhaed, Bdj95, C.Fred, Cyndaquazy, DRIRV19942008, Dabby, DanTD, Fredddie, Freewayguy, GETONERD84, I-10, Jllm06, Juliancolton, KyleAndMelissa22, Lewisistheone1991, Master son, Mechanicsc, Miller17CU94, Morriswa, NE2, Nyttend, OrangeCounty39, Paddu, Proofreader77, Radagast83, Rschen7754, SPUI, Splat5572, TigerKL81, TwinsMetsFan, Woohookitty, 206 anonymous edits

Greenville_County,_South_Carolina *Source*: http://en.wikipedia.org/w/index.php?title=Greenville_County%2C_South_Carolina *Contributors*: Acntx, Akhenaton06, AshyLarry, CanisRufus, D6, Danny, Geoff Plourde, GrahamHardy, Hu12, Inkan1969, JWWalker, Johnpacklambert, Kaldari, Leslie Mateus, Lincolnite, Lisamh, Monegasque, Nyttend, Oldkinderhook, Omnedon, Pearle, Phantomksu, Ram-Man, Roadtimes, Robertjohnsonrj, Scanime, Smallbones, Sonofaque86, Template namespace initialisation script, Thes entinel, Tom.Lineberger, Tstarl0425, Wechselstrom, Wedg, Whhalbert, Whyatt3854, WillC, Worldenc, Xiliquiern, Yahel Guhan, Zosimus, 28 anonymous edits

Hilton_Head_Island,_South_Carolina *Source*: http://en.wikipedia.org/w/index.php?title=Hilton_Head_Island%2C_South_Carolina *Contributors*: 93hotel, A Softer Answer, Absolon, Accurizer, Acherubini2020, Acntx, Aecis, Alex 1331, Alexwcovington, AllenHarkleroad, Allstar86, Amag911, Amandalolik, Andrewnpeters, Backspace, Badagnani, Bdowd, Ben Ben, Bigbackattack, Bluemoose, Bmicomp, Bogey97, Bongwarrior, Booksworm, Brichcja, Bytwerk, Cailin9498, CambridgeBayWeather, Can't sleep, clown will eat me, Carlagolden, Cdamgen, Ckatz, Colonies Chris, CopperSquare, Corpx, Ctbolt, DS1953, Damienpringle, Danilocastro, Darfin35, Darth Mike, Deinocheirus, Demokratickid, Discospinster, Donald Albury, Drbreznjev, Dreadstar, Dryman, Duffman, EdJogg, Epbr123, Excirial, Fishingfoolcool, Fluri, Footie2, Fpo, Func, Fuscob, Gadesignengineer, Golbez, Hhiguy, Hhipete, Hmains, Hotcrocodile, Hu12, Ideal gas equation, Infratec, Iridescent, Irishguy, Itsallthere, JFreeman, Jacuzzi, Jllm06, John Cardinal, Jthomp72, JubJub666, JustAGal, KDills22, Kate, Ken Gallager, Koavf, Kubigula, Lesnail, Lightmouse, Linnell, LongBay, Lukekd, MJCdetroit, MPD01605, Marek69, Mason.Jones, Mecil, Miam, Mike Cline, Moeron, MoodyGroove, MrRadioGuy, Mwanner, NE2, Nagy, NicSIV, Northernpike360, Nposs, Nytimes19992000, Nyttend, Paste, Pfly, Plasticup, Postdlf, Ram-Man, Reader2323, Rearden9, Rich Farmbrough, Richard Harvey, Rjdyork, Rjwilmsi, Rmmiles, Roaddawg, Rodzilla, Roob52, Scplaya69, Seth Ilys, Sevnrock, Skapur, Skosloff, Smalljim, Snowolf, Spellcast, Stupid girl, SunCreator, Tedder, Tenmazero, Tewapack, Thebestaddress, Tony1, Trafton, Trident13, Truck6alpha, Tvscreen, User2004, Usnerd, Vikingboy0129, W4chris, WalkingBill, West.andrew.g, WhisperToMe, Wiffleball, Wilchett, Will Beback, Worker31b, Yassie, 310 anonymous edits

Simpsonville,_South_Carolina *Source*: http://en.wikipedia.org/w/index.php?title=Simpsonville%2C_South_Carolina *Contributors*: Acntx, Almanley, Amstaton, Beaubailey, Bentley4, Bratsche, ColoBill4, Danhash, Degags, E-Kartoffel, Flowanda, Fvhayden, Headinthedoor, Howenstein115, Hubertfarnsworth, Inkan1969, Islandisee, JQLibet, Jllm06, John Cardinal, John K, Jusdafax, Leonard23, Mickey Ocean, Miller17CU94, Milowent, Neurolysis, Nyttend, Ottawa4ever, Pablomartinez, PaulBowley, Ram-Man, Rando99, Readhere, Rjwilmsi, Scmommy, Seth Ilys, Trophyman582, Ultraexactzz, Wikipelli, WillC, 88 anonymous edits

Mauldin,_South_Carolina *Source*: http://en.wikipedia.org/w/index.php?title=Mauldin%2C_South_Carolina *Contributors*: Acntx, BlackNinjaBobby, Chasingsol, Christopher Parham, Da.Tomato.Dude, DandyDan2007, Drewnick, Earthere, Gamecocks85, HKav, Headinthedoor, Hubertfarnsworth, Inkan1969, Islandisee, Jllm06, John K, Johnpacklambert, Ksolo787, Lionclaw, Mboverload, Meco, Musicluver4life, Netparrot, Nyttend, Pearle, Ram-Man, Rich Farmbrough, Seth Ilys, Shoeofdeath, Tedder, Tommy2010, Trvsdrlng, Uviolet, WadeSimMiser, 79 anonymous edits

Image Sources, Licenses and Contributors

File:I-185 (SC).svg *Source*: http://en.wikipedia.org/w/index.php?title=File:I-185_(SC).svg *License*: unknown *Contributors*: Ltljltlj, 2 anonymous edits

File:I-185 (SC) map.svg *Source*: http://en.wikipedia.org/w/index.php?title=File:I-185_(SC)_map.svg *License*: unknown *Contributors*: User:25or6to4

File:I-385 (SC).svg *Source*: http://en.wikipedia.org/w/index.php?title=File:I-385_(SC).svg *License*: unknown *Contributors*: Ltljltlj, 2 anonymous edits

File:US 276.svg *Source*: http://en.wikipedia.org/w/index.php?title=File:US_276.svg *License*: unknown *Contributors*: SPUI, 1 anonymous edits

File:I-85 (SC).svg *Source*: http://en.wikipedia.org/w/index.php?title=File:I-85_(SC).svg *License*: unknown *Contributors*: Ltljltlj, 2 anonymous edits

File:US 25.svg *Source*: http://en.wikipedia.org/w/index.php?title=File:US_25.svg *License*: unknown *Contributors*: Bidgee, SPUI, TwinsMetsFan, Xnatedawgx, 4 anonymous edits

File:US 29.svg *Source*: http://en.wikipedia.org/w/index.php?title=File:US_29.svg *License*: unknown *Contributors*: Bidgee, SPUI, Xnatedawgx, 4 anonymous edits

Image:Greenville, South Carolina 1955 Yellow Book.jpg *Source*: http://en.wikipedia.org/w/index.php?title=File:Greenville,_South_Carolina_1955_Yellow_Book.jpg *License*: unknown *Contributors*: Infrogmation, SPUI, T2

File:South Carolina 417.svg *Source*: http://en.wikipedia.org/w/index.php?title=File:South_Carolina_417.svg *License*: unknown *Contributors*: User:Mr. Matté

File:South Carolina 20.svg *Source*: http://en.wikipedia.org/w/index.php?title=File:South_Carolina_20.svg *License*: unknown *Contributors*: User:Mr. Matté

File:South Carolina 153.svg *Source*: http://en.wikipedia.org/w/index.php?title=File:South_Carolina_153.svg *License*: unknown *Contributors*: User:Mr. Matté

file:SCDOT.svg *Source*: http://en.wikipedia.org/w/index.php?title=File:SCDOT.svg *License*: unknown *Contributors*: Overpush, 1 anonymous edits

File:US 29 map.png *Source*: http://en.wikipedia.org/w/index.php?title=File:US_29_map.png *License*: unknown *Contributors*: w:User:StratosphereNick Nolte

File:US 90.svg *Source*: http://en.wikipedia.org/w/index.php?title=File:US_90.svg *License*: unknown *Contributors*: Bidgee, SPUI, 2 anonymous edits

File:US 98.svg *Source*: http://en.wikipedia.org/w/index.php?title=File:US_98.svg *License*: unknown *Contributors*: Bidgee, SPUI, Xnatedawgx, 2 anonymous edits

File:I-10.svg *Source*: http://en.wikipedia.org/w/index.php?title=File:I-10.svg *License*: unknown *Contributors*: Augiasstallputzer, Infrogmation, Juliancolton, Ltljltlj, Rocket000, SPUI, 1 anonymous edits

File:I-75.svg *Source*: http://en.wikipedia.org/w/index.php?title=File:I-75.svg *License*: unknown *Contributors*: Augiasstallputzer, Ltljltlj, SPUI, 1 anonymous edits

File:I-26 (SC).svg *Source*: http://en.wikipedia.org/w/index.php?title=File:I-26_(SC).svg *License*: unknown *Contributors*: Ltljltlj, 3 anonymous edits

File:I-77.svg *Source*: http://en.wikipedia.org/w/index.php?title=File:I-77.svg *License*: unknown *Contributors*: Augiasstallputzer, Ltljltlj, SPUI, 4 anonymous edits

File:I-40.svg *Source*: http://en.wikipedia.org/w/index.php?title=File:I-40.svg *License*: unknown *Contributors*: Augiasstallputzer, Ltljltlj, SPUI, Xnatedawgx, 1 anonymous edits

File:I-85.svg *Source*: http://en.wikipedia.org/w/index.php?title=File:I-85.svg *License*: unknown *Contributors*: Augiasstallputzer, Ltljltlj, SPUI, 2 anonymous edits

File:I-64.svg *Source*: http://en.wikipedia.org/w/index.php?title=File:I-64.svg *License*: unknown *Contributors*: Augiasstallputzer, Ltljltlj, Rocket000, SPUI, Xnatedawgx

File:I-66.svg *Source*: http://en.wikipedia.org/w/index.php?title=File:I-66.svg *License*: unknown *Contributors*: Augiasstallputzer, Ltljltlj, Rocket000, SPUI, 3 anonymous edits

File:I-70.svg *Source*: http://en.wikipedia.org/w/index.php?title=File:I-70.svg *License*: unknown *Contributors*: , ,

File:MD Route 99.svg *Source*: http://en.wikipedia.org/w/index.php?title=File:MD_Route_99.svg *License*: unknown *Contributors*: Jeff02, Rocket000, Sevela.p

Image:US 29 (FL).svg *Source*: http://en.wikipedia.org/w/index.php?title=File:US_29_(FL).svg *License*: unknown *Contributors*: User:Jeff02

Image:US 170.svg *Source*: http://en.wikipedia.org/w/index.php?title=File:US_170.svg *License*: unknown *Contributors*: SPUI

Image:US blank.svg *Source*: http://en.wikipedia.org/w/index.php?title=File:US_blank.svg *License*: unknown *Contributors*: Fredddie, InterstateKazuya40, Martin H., Rfc1394, SPUI, Sehome Bay, 2 anonymous edits

File:I-385 (SC) map.svg *Source*: http://en.wikipedia.org/w/index.php?title=File:I-385_(SC)_map.svg *License*: unknown *Contributors*: User:25or6to4

File:Business Spur 385.svg *Source*: http://en.wikipedia.org/w/index.php?title=File:Business_Spur_385.svg *License*: unknown *Contributors*: I-215, KelleyCook, Ltljltlj, Mountain169257, SPUI, Sehome Bay, T2, 1 anonymous edits

File:SC Highway 49 exit on I-385.JPG *Source*: http://en.wikipedia.org/w/index.php?title=File:SC_Highway_49_exit_on_I-385.JPG *License*: unknown *Contributors*: User:JamieS93

File:South Carolina 308.svg *Source*: http://en.wikipedia.org/w/index.php?title=File:South_Carolina_308.svg *License*: unknown *Contributors*: User:Mr. Matté

File:South Carolina 49.svg *Source*: http://en.wikipedia.org/w/index.php?title=File:South_Carolina_49.svg *License*: unknown *Contributors*: User:Mr. Matté

File:US 221.svg *Source*: http://en.wikipedia.org/w/index.php?title=File:US_221.svg *License*: unknown *Contributors*: Common Good, SPUI, 2 anonymous edits

File:South Carolina 101.svg *Source*: http://en.wikipedia.org/w/index.php?title=File:South_Carolina_101.svg *License*: unknown *Contributors*: User:Mr. Matté

File:South Carolina 14.svg *Source*: http://en.wikipedia.org/w/index.php?title=File:South_Carolina_14.svg *License*: unknown *Contributors*: User:Mr. Matté

File:South Carolina 418.svg *Source*: http://en.wikipedia.org/w/index.php?title=File:South_Carolina_418.svg *License*: unknown *Contributors*: User:Mr. Matté

File:South Carolina 146.svg *Source*: http://en.wikipedia.org/w/index.php?title=File:South_Carolina_146.svg *License*: unknown *Contributors*: User:Mr. Matté

File:South Carolina 291.svg *Source*: http://en.wikipedia.org/w/index.php?title=File:South_Carolina_291.svg *License*: unknown *Contributors*: User:Mr. Matté

File:Interstate 85 map.png *Source*: http://en.wikipedia.org/w/index.php?title=File:Interstate_85_map.png *License*: unknown *Contributors*: w:User:StratosphereNick Nolte

File:I-65 (AL).svg *Source*: http://en.wikipedia.org/w/index.php?title=File:I-65_(AL).svg *License*: unknown *Contributors*: Ltljltlj, 1 anonymous edits

File:US 82.svg *Source*: http://en.wikipedia.org/w/index.php?title=File:US_82.svg *License*: unknown *Contributors*: Common Good, SPUI, Xnatedawgx

File:Alabama 6.svg *Source*: http://en.wikipedia.org/w/index.php?title=File:Alabama_6.svg *License*: unknown *Contributors*: Ltljltlj, Sevela.p

File:I-20.svg *Source*: http://en.wikipedia.org/w/index.php?title=File:I-20.svg *License*: unknown *Contributors*: Ltljltlj, SPUI, 2 anonymous edits

File:I-95.svg *Source*: http://en.wikipedia.org/w/index.php?title=File:I-95.svg *License*: unknown *Contributors*: Juliancolton, Kazuya35, Ltljltlj, O, SPUI, 3 anonymous edits

File:US 460.svg *Source*: http://en.wikipedia.org/w/index.php?title=File:US_460.svg *License*: unknown *Contributors*: SPUI

Image:I-85construxgeorgia.JPG *Source*: http://en.wikipedia.org/w/index.php?title=File:I-85construxgeorgia.JPG *License*: unknown *Contributors*: User:RandorXeus

Image:Atlanta 75.85.jpg *Source*: http://en.wikipedia.org/w/index.php?title=File:Atlanta_75.85.jpg *License*: unknown *Contributors*: Original uploader was Atlantacitizen at en.wikipedia

Image:SC I-85 North-Exit 1.jpg *Source*: http://en.wikipedia.org/w/index.php?title=File:SC_I-85_North-Exit_1.jpg *License*: unknown *Contributors*: User:Miller17CU94

Image:I-85SouthCarolinaWelcomeSign.jpg *Source*: http://en.wikipedia.org/w/index.php?title=File:I-85SouthCarolinaWelcomeSign.jpg *License*: unknown *Contributors*: Airtuna08 (talk). Original uploader was Airtuna08 at en.wikipedia

Image:I40i85NC.jpg *Source*: http://en.wikipedia.org/w/index.php?title=File:I40i85NC.jpg *License*: unknown *Contributors*: Fietsbel, MPD01605, T2

File:US 278.svg *Source*: http://en.wikipedia.org/w/index.php?title=File:US_278.svg *License*: unknown *Contributors*: SPUI, 2 anonymous edits

File:US 278 map.png *Source*: http://en.wikipedia.org/w/index.php?title=File:US_278_map.png *License*: unknown *Contributors*: User:25or6to4

File:US 59 (1961).svg *Source*: http://en.wikipedia.org/w/index.php?title=File:US_59_(1961).svg *License*: unknown *Contributors*: User:Fredddie, User:SPUI, User:Scott5114

File:US 71 (1961).svg *Source*: http://en.wikipedia.org/w/index.php?title=File:US_71_(1961).svg *License*: unknown *Contributors*: User:Fredddie, User:SPUI, User:Scott5114

File:I-30 (AR) Metric.svg *Source*: http://en.wikipedia.org/w/index.php?title=File:I-30_(AR)_Metric.svg *License*: unknown *Contributors*: Ltljltlj

File:I-55.svg *Source*: http://en.wikipedia.org/w/index.php?title=File:I-55.svg *License*: unknown *Contributors*: Augiasstallputzer, Fran Rogers, Ltljltlj, SPUI, Xnatedawgx, 1 anonymous edits

File:I-22 (AL).svg *Source*: http://en.wikipedia.org/w/index.php?title=File:I-22_(AL).svg *License*: unknown *Contributors*: Ltljltlj

File:I-59 (AL).svg *Source*: http://en.wikipedia.org/w/index.php?title=File:I-59_(AL).svg *License*: unknown *Contributors*: Ltljltlj

File:I-285.svg *Source*: http://en.wikipedia.org/w/index.php?title=File:I-285.svg *License*: unknown *Contributors*: Ltljltlj, SPUI, 2 anonymous edits

File:I-520.svg *Source*: http://en.wikipedia.org/w/index.php?title=File:I-520.svg *License*: unknown *Contributors*: Ltljltlj, SPUI, 3 anonymous edits

File:I-95 (SC).svg *Source*: http://en.wikipedia.org/w/index.php?title=File:I-95_(SC).svg *License*: unknown *Contributors*: Ltljltlj, 2 anonymous edits

File:no image.svg *Source*: http://en.wikipedia.org/w/index.php?title=File:No_image.svg *License*: unknown *Contributors*: SPUI (on English Wikipedia)

File:Business plate.svg *Source*: http://en.wikipedia.org/w/index.php?title=File:Business_plate.svg *License*: unknown *Contributors*: Homefryes, Ltljltlj, RTCNCA, SPUI

Printed by Books on Demand GmbH, Norderstedt / Germany